STATISTICAL MINING AND DATA VISUALIZATION IN ATMOSPHERIC SCIENCES

edited by

Timothy J. Brown
Desert Research Institute, U.S.A.

Paul W. Mielke, Jr.
Colorado State University, U.S.A.

A Special Issue of
Data Mining and Knowledge Discovery
Volume 4, No. 1 (2000)

KLUWER ACADEMIC PUBLISHERS
Boston / Dordrecht / London

DATA MINING AND KNOWLEDGE DISCOVERY

Volume 4, No. 1, April 2000

Special Issue: Statistical Mining and Data Visualization in Atmospheric Sciences
Guest Editors: Timothy J. Brown and Paul W. Mielke, Jr.

Distributors for North, Central and South America:
Kluwer Academic Publishers
101 Philip Drive
Assinippi Park
Norwell, Massachusetts 02061 USA
Telephone (781) 871-6600
Fax (781) 681-9045
E-Mail <kluwer@wkap.com>

Distributors for all other countries:
Kluwer Academic Publishers Group
Distribution Centre
Post Office Box 322
3300 AH Dordrecht, THE NETHERLANDS
Telephone 31 78 6392 392
Fax 31 78 6546 474
E-Mail <services@wkap.nl>

Electronic Services <http://www.wkap.nl>

Library of Congress Cataloging-in-Publication Data

A C.I.P. Catalogue record for this book is available
from the Library of Congress.

Printed on acid-free paper.

Printed in the United States of America

Data Mining and Knowledge Discovery, 4, 5–6 (2000)

Guest Editorial: Statistical Mining and Data Visualization in Atmospheric Sciences

TIMOTHY J. BROWN tbrown@dri.edu
Desert Research Institute

PAUL W. MIELKE, JR. mielke@lamar.colostate.edu
Colorado State University

Data permeates atmospheric science. A full set of sweeps from a NEXRAD-88D radar can record 13 megabytes of data over a five minute period; there are over 120 radar sites across the U.S. At a minimum, 400 megabytes per day of operational gridded numerical weather forecast output are produced at the National Centers for Environmental Prediction. At the National Center for Atmospheric Research in Boulder, Colorado, 150 terabytes of research data are archived on Mass Storage. The National Climatic Data Center maintains over 500 terabytes of global climate records. These are but a few examples of data in the U.S. There are massive volumes of atmospheric empirical observations and model output collected and stored at various operational facilities and research organizations around the world.

With so much data comes the problem of extracting useful information that Knowledge Discovery in Databases attempts to alleviate. The search for patterns, the fitting of models and the exploration of data using visual methods are as relevant to atmospheric science data as any other data intensive discipline. Regardless of the size of the data set, most researchers find themselves using mathematical algorithms and graphics to help them in understanding their data. Similar to other physical, social and health sciences, the high correlation between computing power, database technology and data gathering ability has massively increased the size of atmospheric science databases. Analyses of many of these data require computational intensive methods.

The idea of performing computationally intensive analyses extends back well into the early 1800s. C.F. Gauss (circa 1809) developed linear programming to accomplish the computations required for least sum of absolute deviations regression. Since the lack of computational power prohibited the implementation of linear programming, least sum of squared deviations (least squares) regression was used by Gauss since this alternative approach could be implemented with calculus. When R.A. Fisher developed statistical methods for experimental design in agriculture during the early 1900's, researchers probably were not giving much thought to analyzing terabytes of data. No doubt they would have appreciated the statistical and visualization packages in common use today. Statistical analyses and scientific visualization are now integral parts of atmospheric science.

A large volume of data is not the only reason for the use of computational methods in atmospheric science; the way most of the data are collected guides how the analysis should proceed. Examples of experimental design in atmospheric science are few compared to other disciplines such as the social and health sciences. Many traditional statistical procedures, as found in standard introductory statistics textbooks, must be applied with care, or perhaps even abandoned in the atmospheric sciences. The dynamic nature of the atmospheric system prohibits "repeated measurements" in the sense of classical statistics. The assessment of chance outcome must then be formulated via alternative non-classical methods, which generally are highly computational because they require only a very few reasonable assumptions such as randomization.

In atmospheric science, as well as any other discipline, the visualization of data should not only be an acceptable practice, but a necessary one. J. Tukey and colleagues (circa 1980) brought to the forefront the concepts of exploratory data analysis. The idea simply asks investigators to look at their data. Computational and graphics technology has advanced to a level where one can easily examine a single time series or generate interactive displays of multivariate and multidimensional data.

The papers in this special issue have evolved from a National Science Foundation sponsored workshop held in 1996 called "Computer Intensive Statistics and Applications in the Atmospheric Sciences". This workshop was held in conjunction with the American Meteorological Society 13th Conference on Probability and Statistics in the Atmospheric Sciences. A number of prominent statisticians were invited to give lectures on a variety of topics including resampling methods, graphical displays, data mining and visualization. Four papers are presented in this issue, covering bootstrap and permutation methods, visualization of large multivariate data sets, and the visualization of spatial data and map displays. These papers represent only a subset of data mining topics that could be discussed in relation to the field of atmospheric science. However, we hope that they will give the reader an introduction and insight to the possibilities of knowledge discovery in atmospheric science.

We would like to thank the editors of the Data Mining and Knowledge Discovery, especially Usama Fayyad, for allowing us the opportunity to put this special issue together. We would also like to thank the authors in this special issue for their contributions to this new and exciting field of discovery.

Timothy J. Brown is a research scientist at the Desert Research Institute, and graduate faculty in the Atmospheric Sciences Program, University of Nevada. He received his Ph.D. in Climatology from the University of Colorado in 1995. His research interests include the application of statistical methods and visualization techniques to atmospheric science data.

Paul W. Mielke, Jr. is a professor of Statistics at the Colorado State University. He received a Ph.D. degree from the University of Minnesota in 1963. His research interests include permutation techniques, contingency table analyses, weather modification verifications, and short term climatic predictions.

Data Mining and Knowledge Discovery, 4, 7–27 (2000)

Euclidean Distance Based Permutation Methods in Atmospheric Science

PAUL W. MIELKE, JR. mielke@lamar.colostate.edu
Department of Statistics, Colorado State University, Fort Collins, CO 80523-1877, USA

KENNETH J. BERRY berry@lamar.colostate.edu
Department of Sociology, Colorado State University, Fort Collins, CO 80523-1784, USA

Editors: Timothy Brown and Paul Mielke, Jr.

Abstract. The majority of existing statistical methods inherently involve complex nonmetric analysis spaces due to their least squares regression origin; consequently, the analysis space of such statistical methods is not consistent with the simple metric Euclidean geometry of the data space in question. The statistical methods presented in this paper are consistent with the data spaces in question. These alternative methods depend on exact and approximate permutation procedures for univariate and multivariate data involving cyclic phenomena, autoregressive patterns, covariate residual analyses including most linear model based experimental designs, and linear and nonlinear prediction model evaluations. Specific atmospheric science applications include climate change, Atlantic basin seasonal tropical cyclone predictions, analyses of weather modification experiments, and numerical model evaluations for phenomena such as cumulus clouds, clear-sky surface energy budgets, and mesoscale atmospheric predictions.

Keywords: agreement, autoregressive patterns, cyclic data, distribution-free, experimental designs, inference, multivariate, nonparametric, permutation, prediction, regression, residual analyses

1. Introduction

While statistical analyses were originally based on between-value Euclidean distances, this approach changed around 1800 with the necessity for complex regression analyses. Least sum of Absolute Deviations (LAD) regression appears in the early works of Boscovich (1757), Laplace (1789), and Gauss (1809). Sheynin (1973) documents that Gauss developed linear programming to obtain the linear model parameters of LAD regression and, because obtaining the linear model parameters was too computationally intensive prior to the development of computers, Gauss used Least sum of Squared Deviations (LSD, also termed least squares) regression by default since calculus easily yielded the linear model parameters of LSD regression. The American mathematician and astronomer Bowditch (1809) argued for the use of LAD regression over LSD regression because squared prediction errors unduly emphasize extreme observations (Sheynin, 1973). The dependence of classical statistical techniques such as analysis of variance and analysis of covariance on squared Euclidean distances is an immediate consequence of LSD regression. The problem with the use of squared Euclidean distance is that it does not satisfy the triangle inequality of a metric (Mielke, 1987; Rudin, 1966, p. 9). Consider the values 3, 4 and 5 on the real number line. The Euclidean distance from 3 to 5 is equal to the sum of the Euclidean distances from 3 to 4 and 4 to 5. However, the squared Euclidean distance from 3 to 5 is greater than the sum of squared Euclidean distances from 3 to 4 and 4 to 5, i.e., $4 > 1 + 1$ which contradicts the triangle inequality. Since the replacement of Euclidean distance with squared Euclidean distance violates the triangle inequality, statistical techniques based on squared Euclidean distances may yield counterintuitive results (Mielke, 1991; Mielke and Berry, 1994).

The goal of this paper is to present permutation-based inference methods applicable to atmospheric science research which are consistent with Euclidean geometry and avoid assumptions such as normality which rarely are met in practice. To achieve this goal, two permutation techniques termed Multi-Response Permutation Procedures (MRPP) and Multi-response Randomized Block permutation Procedures (MRBP) are utilized (Mielke, 1991). Both MRPP and MRBP depend on the concept of randomization due to Fisher (1935) and the choice of distance function which is consequently restricted to Euclidean distance. Since the data encountered in either atmospheric science or most any other branch of science are perceived in a metric Euclidean space (certainly not a nonmetric geometry based on squared Euclidean distance), the recommended statistical methods of this paper operate in analysis spaces which are consistent with the data spaces in question (Mielke, 1984, 1985, 1986, 1991).

2. MRPP

The applications in this section are based on Multi-Response Permutation Procedures (MRPP). Following an initial description of MRPP, a variety of applications are described which yield distribution-free analyses for several types of data routinely encountered in atmospheric science research.

Let $\Omega = \{\omega_1, \ldots, \omega_N\}$ represent a finite collection (sample) of objects which are drawn from a target population of interest. Let $x'_I = (x_{1I}, \ldots, x_{rI})$ denote r commensurate response measurements for object ω_I $(I = 1, \ldots, N)$ and let $S_1, \ldots, S_{g+1}$ designate an exhaustive partitioning of the N objects comprising Ω into $g + 1$ disjoint groups. Also let $\Delta_{I,J}$ be a symmetric distance function value of the response measurements associated with objects ω_I and ω_J. The MRPP statistic is given by

$$\delta = \sum_{i=1}^{g} C_i \xi_i \tag{2.1}$$

where

$$\xi_i = \binom{n_i}{2}^{-1} \sum_{I<J} \Delta_{I,J} \Psi_i(\omega_I) \Psi_i(\omega_J) \tag{2.2}$$

is the average distance function value for all distinct pairs of objects in groups S_i $(i = 1, \ldots, g)$, $n_i \geq 2$ is the number of a priori objects classified into group S_i $(i = 1, \ldots, g)$, $K = \sum_{i=1}^{g} n_i$, $n_{g+1} = N - K \geq 0$ is the number of remaining (i.e., unclassified) objects in the excess group S_{g+1} which is an empty group in many applications, $\sum_{I<J}$ is the sum over all I and J such that $1 \leq I < J \leq N$, $\Psi_i(\omega_I)$ is 1 if ω_I belongs to S_i and 0 otherwise, $C_i > 0$ $(i = 1, \ldots, g)$, and $\sum_{i=1}^{g} C_i = 1$. The null hypothesis (H_0) states that equal probabilities are assigned to each of the

$$M = \frac{N!}{\prod_{i=1}^{g+1} n_i!} \tag{2.3}$$

possible allocations of the N objects to the $g + 1$ groups. Thus the δ compares the within-group clumping of response measurements against the model specified by the random allocation under H_0. The initial presentation and subsequent development of MRPP are given in Brockwell et al. (1982), Brown (1982), Denker and Puri (1988), Mielke (1978, 1979, 1984, 1985, 1986, 1991), Mielke et al. (1976, 1981a), O'Reilly and Mielke (1980), and Robinson (1983). It should be emphasized that the excess group (S_{g+1}) is not merely a mathematical construct. For an example application, see Mielke et al. (1983) where the excess group was essential to an environmental analysis involving the geographical concentration of lead in an urban area.

The choice of the classified group weights $(C_1, \ldots, C_g)$ and the symmetric distance function $\Delta_{I,J}$ specifies the version of MRPP to be employed. While $C_i = n_i/K$ $(i = 1, \ldots, g)$ is the recommended group weighting since it is associated with efficient versions of MRPP, other choices of group weights such as $C_i = (n_i - 1)/(K - g)$, $C_i = 1/g$, and $C_i = n_i(n_i - 1)/\sum_{j=1}^{g} n_j(n_j - 1)$ have also been considered (Mielke, 1984). In most applications of MRPP the symmetric distance function is given by

$$\Delta_{I,J} = \begin{cases} d_{I,J} & \text{if } d_{I,J} \leq B \\ B & \text{otherwise} \end{cases} \tag{2.4}$$

where

$$d_{I,J} = \left[\sum_{h=1}^{r} (x_{hI} - x_{hJ})^2 \right]^{\frac{v}{2}}, \tag{2.5}$$

$B > 0$ is a specified truncation constant, and $v > 0$ is a specified power constant (note that $\Delta_{I,J}$ is ordinary Euclidean distance when B is ∞ and $v = 1$). The choice of B is purely subjective since its use includes the detection of multiple clumping among objects of a single group. When multiple clumping exists, a small B mitigates the between clump distances and emphasizes the within clump distances (Mielke, 1991). Although $v = 2$ yields a nonmetric distance function since the triangle inequality is not satisfied (Mielke, 1986), the choice of $v = 2$ is associated with many commonly used statistical methods, e.g., the two-sample t test and all analysis of variance procedures derived from least squares regression. Mielke (1986, 1991) and Mielke and Berry (1994) provide simple examples demonstrating that statistical methods based on $v = 2$ yield peculiar counterintuitive results when compared to statistical methods based on $v = 1$ for the same data sets. This occurs because the geometry of analyses involving methods based on $v = 2$ differs from the underlying geometry of the data in question. However, the direct correspondence between the geometries of the data and analyses based on $v = 1$ allows intuition to play a major role in constructing special purpose permutation methods. In addition, note that when $v = 1$ MRPP is a median based technique, but when $v = 2$ MRPP is a mean based technique. For clarification, consider the pairwise sum of univariate (i.e., $r = 1$) symmetric distance functions given by

$$\sum_{i<j} \Delta(x_i, x_j) = \sum_{i<j} |x_i - x_j|^v \tag{2.6}$$

where $x_1, \ldots, x_m$ are univariate response variables and $\sum_{i<j}$ is the sum over all i and j such that $1 \le i < j \le m$. Let $x_{1,m} \le \cdots \le x_{m,m}$ be the order statistics associated with $x_1, \ldots, x_m$. If $v = 1$, then the inequality given by

$$\sum_{i=1}^{m} |m - 2i + 1||x_{i,m} - \theta| \ge \sum_{i<j} |x_i - x_j| \tag{2.7}$$

holds for all θ and equality holds if θ is the median of $x_i, \ldots, x_m$. If $v = 2$, then the inequality given by

$$m \sum_{i=1}^{m} (x_i - \theta)^2 \ge \sum_{i<j} (x_i - x_j)^2 \tag{2.8}$$

holds for all θ and equality holds if θ is the mean of $x_1, \ldots, x_m$.

The P-value associated with an observed value of δ (say, δ_o) is the probability under the null hypothesis (H_0) of observing a value of δ as extreme or more extreme than δ_o. Thus, depending on the research question specified, an exact P-value may be expressed as

$$P(\delta \le \delta_o \mid H_0) = \frac{\text{number of } \delta\text{'s} \le \delta_o}{M} \tag{2.9}$$

or

$$P(\delta \ge \delta_o \mid H_0) = \frac{\text{number of } \delta\text{'s} \ge \delta_o}{M}. \tag{2.10}$$

Although an efficient algorithm exists for computing exact MRPP P-values (Berry and Mielke, 1984), when M is large it is impractical to generate the discrete probability distribution of δ. It then becomes necessary to approximate the distribution of δ by a continuous distribution in order to determine the P-value for an observed value of δ. For example, if $g = 3$, $N = 30$, and $n_1 = n_2 = n_3 = 10$, $M = 5.55 \times 10^{12}$. In these cases, a Pearson type III distribution P-value approximation is used which depends on the exact mean, variance, and skewness of δ under H_0 given by

$$\mu = \frac{1}{M}\sum_{I=1}^{M}\delta_I, \tag{2.11}$$

$$\sigma^2 = \frac{1}{M}\sum_{I=1}^{M}(\delta_I - \mu)^2, \tag{2.12}$$

and

$$\gamma = \frac{\frac{1}{M}\sum_{I=1}^{M}(\delta_I - \mu)^3}{\sigma^3}, \tag{2.13}$$

respectively. In particular, the standardized statistic given by

$$T = \frac{\delta - \mu}{\sigma} \tag{2.14}$$

is presumed to follow the Pearson type III distribution (a specific gamma distribution) with the density function given by either

$$f(y) = \frac{(-2/\gamma)^{4/\gamma^2}}{\Gamma(4/\gamma^2)}\,[-(2+y\gamma)/\gamma]^{(4-\gamma^2)/\gamma^2}\,e^{-2(2+y\gamma)/\gamma^2} \tag{2.15}$$

where $-\infty < y < -2/\gamma$ and $\gamma < 0$ or

$$f(y) = \frac{(2/\gamma)^{4/\gamma^2}}{\Gamma(4/\gamma^2)}\,[(2+y\gamma)/\gamma]^{(4-\gamma^2)/\gamma^2}\,e^{-2(2+y\gamma)/\gamma^2} \tag{2.16}$$

where $-2/\gamma < y < \infty$ and $\gamma \geq 0$. It should be noted that

$$f(y) = \frac{1}{\sqrt{2\pi}}e^{-\frac{y^2}{2}} \tag{2.17}$$

if $\gamma = 0$, i.e., the standard normal distribution. If the observed value of T is

$$T_o = \frac{\delta_o - \mu}{\sigma}, \tag{2.18}$$

then

$$P(\delta \leq \delta_o \mid H_0) \doteq \int_{-\infty}^{T_o} f(y)\,dy \tag{2.19}$$

and

$$P(\delta \geq \delta_o \mid H_0) \doteq \int_{T_o}^{\infty} f(y)\, dy \tag{2.20}$$

denote approximate P-values which are evaluated numerically using, say, Simpson's rule over an appropriate finite interval. The Pearson type III distribution is used to approximate the permutation distribution of T because it is completely specified by γ (the skewness of T) and because it includes the normal and chi-squared distributions as special cases. Efficient computational expressions for μ, σ^2, and γ under H_0 are given by Mielke (1991) and Mielke et al. (1976).

Two major advantages that P-value approximations based on exact moments have over simulated (also termed Monte Carlo, rerandomized, or bootstrap) P-value approximations are that they avoid the possibility of an additional type I statistical error due to the simulation process and that quantitative P-value approximations are possible even when the P-values are very small. A concern with Monte Carlo P-value approximations is the amount of type I error compounded by the sampling of all possible permutations. When a with replacement sample of permutations is obtained, an additional type I error results because the P-value now depends on both the observed δ and the Monte Carlo sample of δ values. A more serious concern is that an investigator can draw repeated Monte Carlo samples of δ values until a sample suits the investigator (Berry and Mielke, 1983; Mielke and Berry, 1994; Mielke et al., 1981b). An immediate application of P-values is the classification of an additional object to one of the $g + 1$ disjoint groups $(S_1, \ldots, S_{g+1})$. The selected group yields the smallest P-value obtained among the $g + 1$ P-values when the additional object is appended (if the additional object is classified into group S_{g+1}, then it is an unclassified object).

In addition to the previously discussed probabilistic interpretations involving P-values, it is sometimes of interest to have a measure of the agreement among object measurements within the classified groups. Such a measure for this purpose is the chance-corrected within-group agreement measure given by

$$\mathfrak{R} = 1 - \frac{\delta}{\mu} \tag{2.21}$$

(Berry and Mielke, 1992). Since μ is a constant under H_0, the permutation distributions of δ and $\mathfrak{R}$ are equivalent, i.e.,

$$P(\delta \leq \delta_o \mid H_0) = P\left(\mathfrak{R} \geq 1 - \frac{\delta_o}{\mu} \,\middle|\, H_0\right). \tag{2.22}$$

The values of $\mathfrak{R}$ range from negative values to a maximum of 1 for the extreme case when all object measurements within each classified group are identical (i.e., $\delta = 0$). Since the mean of $\mathfrak{R}$ under H_0 is 0, homogeneity of within-classified group object measurements is associated with $\mathfrak{R} > 0$ and heterogeneity of within-classified group object measurements is associated with $\mathfrak{R} < 0$. The distribution of $\mathfrak{R}$ is usually asymmetric and the upper and lower bounds depend on both the nature of the data and the structure of δ. The degree of homogeneity or heterogeneity depends on the permutation distribution of $\Re$. If large values of $n_1, \ldots, n_g$ and N are involved, a very small value of $P(\delta \leq \delta_o \mid H_0)$ may be associated with an observed small positive value of $\Re$, say $\Re_o$. Similarly, a large value of $\Re_o$ may be associated with a relatively large value of $P(\delta \leq \delta_o \mid H_0)$ if small values of $n_1, \ldots, n_g$ and N are involved. Incidentally, the unbiased correlation ratio is a special case of $\Re$ when $v = 2$, $C_i = (n_i - 1)/(K - g)$ and $K = N$ (Kelley, 1935).

While only the MRPP with $v = 1$ is consistent with Euclidean geometry, MRPP includes very broad classes of permutation tests. In particular, MRPP with $v = 2$ includes many classical permutation tests such as one-way analysis of variance, Hotelling's T^2, and the Bartlett-Nanda-Pillai trace (Mielke and Berry, 1994). Simulated power comparisons for location shifts of five uncorrelated bivariate distributions (Mielke and Berry, 1994) indicate that (1) MRPP with either Euclidean commensuration, Hotelling commensuration and $v = 1$, or Hotelling commensuration and $v = 2$ performs almost equally well for the bivariate normal distribution, (2) MRPP with Hotelling commensuration and $v = 2$ performs better than either MRPP with Euclidean commensuration or Hotelling commensuration and $v = 1$ for the bivariate uniform distribution, and (3) either MRPP with Euclidean commensuration or Hotelling commensuration and $v = 1$ performs much better than MRPP with Hotelling commensuration and $v = 2$ for the bivariate exponential, lognormal and Cauchy distributions. A related study involved simulated power comparisons for location shifts of five correlated bivariate distributions (Mielke and Berry, 1999). If the location shift is parallel to the major axis of a correlated bivariate distribution, then MRPP with Euclidean commensuration performs better than either MRPP with Hotelling commensuration and $v = 1$ or MRPP with Hotelling commensuration and $v = 2$ for all five correlated bivariate distributions. If the location shift is parallel to the minor axis of a correlated bivariate distribution, then MRPP with Hotelling commensuration and $v = 2$ performs better than the other two versions of MRPP for the correlated bivariate normal and uniform distributions. Finally, if the location shift is parallel to the minor axis of a correlated bivariate distribution, then MRPP with Hotelling commensuration and $v = 1$ performs better than the other two versions of MRPP for the correlated bivariate exponential, lognormal, and Cauchy distributions.

2.1. *P-value comparisons of univariate MRPP with $v = 1$ and $v = 2$*

Following an example in Mielke and Berry (1994) to demonstrate the stability of MRPP with $v = 1$, consider the data given in Table 1 where $N = 26$, $g = 2$, $r = 1$, $n_1 = n_2 = 13$, and the truncation constant B of expression (2.4) is ∞. While 25 of the $N = 26$ values are fixed, one value of group S_1, denoted by z, is allowed to vary in order to determine its effect on the exact two-sided P-values. For this example $C_1 = C_2 = 0.5$ in both MRPP δ statistics since the four choices of C_i are equivalent. If $A_1 = \sum_{I=1}^{N} x_{1I}$ and $A_2 = \sum_{I=1}^{N} x_{1I}^2$, then the relationship between the two-sample t test statistic, say t, and the MRPP δ statistic with $v = 2$ is

$$N\delta = 2(NA_2 - A_1^2)/(N - 1 + t^2) \tag{2.23}$$

Table 1. The observed value frequencies of objects in groups S_1 and S_2 where $N = 26$ and $n_1 = n_2 = 13$.

Values	S_1	S_2
405.6	1	0
405.7	2	0
405.8	4	0
405.9	4	1
406.0	1	3
406.1	0	4
406.2	0	3
406.3	0	2
z	1	0

Table 2. Exact two-sided P-value comparison of MRPP with $v = 1$ and $v = 2$ together with the two-sample t test for the data in Table 1.

	MRPP		Two-sample
z	$v = 1$	$v = 2$	t test
67.7	4.04×10^{-6}	4.04×10^{-6}	0.322
397.7	4.04×10^{-6}	4.04×10^{-6}	0.153
403.7	4.04×10^{-6}	4.04×10^{-6}	1.16×10^{-2}
404.7	4.04×10^{-6}	4.04×10^{-6}	5.69×10^{-4}
405.7	4.04×10^{-6}	4.04×10^{-6}	5.84×10^{-7}
406.7	9.81×10^{-5}	9.82×10^{-3}	9.16×10^{-3}
407.7	9.81×10^{-5}	0.430	0.320
409.7	9.81×10^{-5}	1.000	1.000
413.7	9.81×10^{-5}	1.000	0.617
743.7	9.81×10^{-5}	1.000	0.333

(Mielke and Berry, 1994) where A_1 and A_2 are invariant under H_0 of MRPP. Based on the data of Table 1, the exact two-sided P-value comparisons of MRPP with $v = 1$ and $v = 2$ together with the two-sample t test (assuming independence, common variance, and normality) are contained in Table 2 for selected values of z.

With respect to the 25 fixed values, the MRPP P-values with $v = 1$ for low and high values of z are 4.04×10^{-6} and 9.81×10^{-5}, respectively. In contrast, the MRPP P-values with $v = 2$ are 4.04×10^{-6} for low values of z and 1.00 for high values of z. While a single extreme value can overwhelm MRPP with $v = 2$, a single extreme value cannot overwhelm MRPP with $v = 1$. As a consequence of expression (2.23), the P-value differences between MRPP with $v = 2$ and the two-sample t test correspond to the nonparametric and parametric assumptions, respectively. Since the limiting value of t^2 approaches 1 as z becomes either very low or very high (Mielke and Berry, 1994), the P-value of the two-sample t test approaches 0.327; this result is consistent with the two-sample t test P-values in Table 2.

The permutation approach provides the convenient implementation of intuitively satisfying statistical techniques. The continued blind use of counterintuitive paradigms such as squared Euclidean distance based statistical techniques, with or without distributional assumptions, is absurd.

2.2. *Analyses involving cyclic data*

Suppose a study is concerned with comparing groups of points on the surface of an m-dimensional hypersphere. If x_I and x_J denote two of N points on the surface of a unit m-dimensional hypersphere centered at the origin, then $\sum_{k=1}^{m} x_{kI}^2 = 1$ for $I = 1, \ldots, N$ and the symmetric distance measure is the arc length in radians given by

$$\Delta_{I,J} = \cos^{-1}\left(\sum_{k=1}^{m} x_{kI} x_{kJ}\right) . \tag{2.24}$$

If $m = 2$, let θ_I and θ_J denote the positions in radians of the Ith and Jth points on the edge of a unit circle ($0 \leq \theta_I < 2\pi$ for $I = 1, \ldots, N$). Then the symmetric distance measure is alternatively given by

$$\Delta_{I,J} = \min(|\theta_I - \theta_J| , 2\pi - |\theta_I - \theta_J|) . \tag{2.25}$$

This representation is especially convenient for group comparisons involving cyclic data such as wind directions or temporal occurrence patterns during either days or years. For example, if the Ith event occurs h_I days after the first day of a year, then $0 \leq h_I < 365$ and $\theta_I = 2\pi h_I / 365$. Furthermore, suppose $\theta_{1I}, \ldots, \theta_{cI}$ denotes c multi-response cyclic measurements in radians of the Ith of N objects. Then the transformation given by

$$\begin{aligned} x_{1I} &= \cos(\theta_{1I}), \\ x_{2I} &= \sin(\theta_{1I}) \cos(\theta_{2I}), \\ &\vdots \\ x_{cI} &= \sin(\theta_{1I}) \sin(\theta_{2I}), \ldots, \sin(\theta_{c-1,I}) \cos(\theta_{cI}), \\ x_{c+1,I} &= \sin(\theta_{1I}) \sin(\theta_{2I}), \ldots, \sin(\theta_{c-1,I}) \sin(\theta_{cI}), \end{aligned} \tag{2.26}$$

yields the basis for MRPP analyses of c-dimensional cyclic data since, in the present context, $m = c + 1$ and $x_{1I}, \ldots, x_{c+1,I}$ denote the coordinates of a point on a unit hypersphere having $c + 1$ dimensions (Mielke, 1986).

Similarly, suppose groups involving both wind direction and velocity are compared. Let θ_I and υ_I be the polar coordinates denoting direction in radians and velocity for the Ith of N measurements, respectively. Then, $x_{1I} = \upsilon_I \cos(\theta_I)$ and $x_{2I} = \upsilon_I \sin(\theta_I)$ are the corresponding Cartesian coordinates. Here the symmetric distance measure ($\Delta_{I,J}$) of MRPP analyses is again the Euclidean distance based on the Ith and Jth Cartesian coordinates.

2.3. *First and higher order autoregressive pattern detection methods*

Consider the variation of MRPP where $r \geq 1$, $v = 1$, $g = N$, B is ∞, $n_i = 1$ $(i = 1, \ldots, g)$, a meaningful order exists for the objects $\omega_1, \ldots, \omega_N$ (e.g., occurrences in time), $\Delta_{I,J}$ is defined in expression (2.4), and $M = N!$. Then the variation of the MRPP statistic given by

$$\delta = \frac{1}{N-1} \sum_{I=2}^{N} \Delta_{I-1,I} \tag{2.27}$$

suggests first-order autoregressive occurrences when δ is small. Thus, the P-value $= P(\delta \leq \delta_o \mid H_0)$ and efficient computational expressions for the exact mean, variance, and skewness of δ under H_0 to obtain a Pearson type III distribution P-value approximation are given by Mielke (1991) and Walker et al. (1997). Incidentally, if $r = 1$ and $v = 2$, it is noted that this approach yields a permutation version of the Durbin and Watson (1950) test for serial correlation.

Suppose $N = ij + k$ where i, j, and k are integers such that $i \geq 1$, $j \geq 1$, and $0 \leq k < i$. Then the jth order autoregressive pattern of this finite sequence is defined as

$$\omega_1, \omega_{j+1}, \omega_{2j+1}, \ldots, \omega_{j+2}, \omega_{2j+2}, \ldots, \omega_j, \ldots, \omega_{ij} \tag{2.28}$$

and each partial sequence of objects (e.g., ω_2, ω_{j+2}, $\omega_{2j+2}, \ldots$) is completed before returning to the beginning of the sequence. Thus, $j - 1$ returns will occur for the jth order autoregressive pattern (N is fairly large and j is quite small in most applications). In the present context, the analysis of the jth order autoregressive pattern involves replacement of the 1st order autoregressive pattern with the jth order autoregressive pattern. While the $j - 1$ returns may reduce the sensitivity of the test for detecting an alternative hypothesis, the permutation test is perfectly valid under H_0.

2.4. *Covariate analyses of residuals*

Consider the multiple regression model given by

$$y_i = \sum_{j=1}^{s} x_{ij} \beta_j + e_i \tag{2.29}$$

where y_i represents the ith of N responses, possibly affected by a treatment; x_{ij} is the jth of s covariates associated with the ith response, where $x_{ij} = 1$ if the model includes an intercept; β_j denotes the jth of s regression parameters; and e_i designates the error associated with the ith response. If the estimates of $\beta_1, \ldots, \beta_s$ which minimize

$$\sum_{i=1}^{N} |e_i| \tag{2.30}$$

are denoted by $\tilde{\beta}_1, \ldots, \tilde{\beta}_s$, then the N residuals of the Least sum of Absolute Deviations (LAD) regression model are given by

$$\tilde{e}_i = y_i - \sum_{j=1}^{s} x_{ij}\tilde{\beta}_j \tag{2.31}$$

for $i = 1, \ldots, N$. An algorithm for obtaining $\tilde{\beta}_1, \ldots, \tilde{\beta}_s$ is given by Barrodale and Roberts (1973, 1974). Portnoy and Koenker (1997) discuss algorithms for large data sets which may be more efficient than least squares regression.

It is widely recognized that multiple linear regression can be used to analyze the results of analysis of variance experimental designs, given a design matrix incorporating an appropriate coding scheme. Consequently, multiple linear regression is often the statistical procedure of choice for anlayzing the results of a wide variety of research experiments. In a two-step process, a linear model is created in which the main or interaction effect of interest is removed from a full model and the reduced model residuals are generated by LAD regression. Second, the residuals are arranged into treatment groups and analyzed by MRPP. The combination of LAD regression and MRPP provides a distribution-free statistical test which is highly resistant to extreme values. Potential applications of MRPP to LAD regression residuals consist of, but are not limited to, completely randomized and randomized block experiments, including one-way, factorial, analysis of covariance, split-plot, and Latin square designs.

Let the N residuals $(\tilde{e}_1, \ldots, \tilde{e}_N)$ be associated with g distinct treatments where n_i residuals correspond to the ith treatment $(i = 1, \ldots, g)$. Then MRPP with $r = 1,\ v = 1,\ B$ is $\infty,\ \Delta_{I,J}$ is defined in expression (2.4), $C_i = n_i/K$ and $n_i \geq 2$ $(i = 1, \ldots, g)$ yields a permutation test for analyzing differences among the g groups of residuals associated with the g treatments. Thus LAD regression coupled with MRPP provides analyses of linear model residuals which are consistent with Euclidean geometry.

Previously described restrictive residual analyses which are consistent with Euclidean geometry involved weather modification studies (Mielke et al., 1982; Smith et al., 1997; Wong et al., 1983). A more general description of residual analyses consistent with Euclidean geometry encompasses balanced and unbalanced experimental designs (Mielke and Berry, 1997a). In addition, a distribution-free single-parameter tolerance interval method consistent with Euclidean geometry is discussed by Smith et al. (1997).

3. MRBP

Suppose a number of observed fields are compared to corresponding fields generated by one or more numerical models. Let the observed phenomena and the one or more numerical model predictions of these phenomena be termed blocks (i.e., the first block represents the observed phenomena and the remaining $b - 1$ blocks represent the $b - 1$ numerical model predictions of these phenomena for a total of b blocks). Also let r denote the number of commensurate response measurements (vector size) from each phenomenon and let g denote the number of phenomena (which will be termed treatments) in question. Suppose $x'_{ij} = (x_{1ij}, \ldots, x_{rij})$ is the number of r response measurements associated with the ith treatment and jth block. Then, the Multi-response Randomized Block permutation Procedures (MRBP) statistic is given by

$$\delta = \left[g \binom{b}{2} \right]^{-1} \sum_{i=1}^{g} \sum_{j<k} \Delta(x_{ij}, x_{ik}) \tag{3.1}$$

where $\sum_{j<k}$ is the sum over all j and k such that $1 \leq j < k \leq b$ and $\Delta(x, y)$ is the symmetric distance function value of the two points (r-dimensional vectors) $x' = (x_1, \ldots, x_r)$ and $y' = (y_1, \ldots, y_r)$ in the r-dimensional Euclidean space. The choice of symmetric distance functions considered here is given by

$$\Delta(x, y) = \left[\sum_{i=1}^{r} (x_i - y_i)^2 \right]^{\frac{v}{2}} \tag{3.2}$$

where $v > 0$ (note that $v = 1$ yields ordinary Euclidean distance). While a truncation constant corresponding to expression (2.4) is omitted from expression (3.2), some applications may find the truncation constant to be very useful. The null hypothesis (H_0) states that the distribution of δ assigns an equal probability to each of the

$$M = (g!)^b \tag{3.3}$$

possible allocations of the g r-dimensional response measurements to the g treatment positions within each of the b blocks. Thus the δ statistic compares the within-group clumping of response measurements against the model specified by the random allocation under H_0. The initial presentations and theoretical developments of MRBP are given in Brockwell and Mielke (1984), Brown (1982), Mielke (1984, 1986), Mielke and Berry (1982), and Mielke and Iyer (1982).

The exact MRBP P-values are analogous to the exact MRPP P-values described in Section 2. Like MRPP, the calculation of exact MRBP P-values becomes unreasonable when M of expression (3.3) exceeds, say, 10^7 (e.g., $b = 6$ and $g = 5$ yield $M = 2.99 \times 10^{12}$, and $b = 3$ and $g = 8$ yield $M = 6.55 \times 10^{13}$). Thus, as with MRPP, a Pearson type III distribution P-value approximation is used which depends on the exact mean, variance, and skewness of δ under H_0. As a result, the Pearson type III distribution approach described in Section 2 is equally appropriate here. Efficient computational expressions for μ, σ^2, and γ are contained in Mielke (1991) and Mielke and Iyer (1982).

The chance-corrected agreement measure given by

$$\Re = 1 - \frac{\delta}{\mu} \tag{3.4}$$

estimates the composite measurement of agreement between blocks for all treatments. As an example, perfect agreement between the observed data and the results of a numerical model would imply that the observed and model predicted data sets (blocks) are identical for each event among the number of events (treatments) in question. As with the chance-corrected within-group agreement measure in expression (2.21), the values of $\Re$ range from negative values to a maximum of 1 when perfect agreement is achieved, the mean of $\Re$ under H_0 is 0, and either agreement or disagreement is implied by $\Re > 0$ or $\Re < 0$, respectively. As noted in Section 2, large data sets can yield small positive values of $\Re$ with very small P-values and small data sets can yield relatively large positive values of $\Re$ with fairly large P-values. While P-values are highly dependent on sample size, this sample size dependency does not hold for the agreement measure $\Re$. Incidentally, special cases of $\Re$ include the classical Pearson product-moment correlation coefficient (which differs by at most a constant multiple) and Spearman's rho (Mielke, 1984). While $\Re$ was initially suggested by Mielke (1984) and termed ρ, a more recent discussion by Berry and Mielke (1988) demonstrates that $\Re$ includes a wide variety of agreement measures as special cases.

The choice of $\Re$ rather than the Pearson product-moment correlation coefficient (r) is that the latter measures linearity and not agreement (Mielke et al., 1996, 1997; Willmott, 1981, 1982). Because of the problems with r, many researchers have turned to measures of agreement in assessing prediction accuracy (Cotton et al., 1994; Elsner and Schmertmann, 1993; Gray et al., 1992; Hess and Elsner, 1994; Kelly et al., 1989; Lee et al., 1995; Tucker; et al., 1989; Willmott, 1981, 1982; Willmott et al., 1985). Watterson (1996) provides a comprehensive comparison of measures of agreement.

A major point noted by Mielke and Iyer (1982) involves aligning blocks prior to an MRBP analysis. For example, if the response measurement medians of each block are subtracted from the corresponding response measurements of each block, then the adjusted response measurement medians are said to be aligned to zero for all blocks. Consider the following univariate data ($r = 1$) where $b = 2$ and $g = 3$: $x_{111} = 4$, $x_{112} = 2$, $x_{113} = 3$, $x_{121} = 9$, $x_{122} = 7$, and $x_{123} = 8$. If $v = 1$, then $\delta_o = 5$ and the statistic δ is 5 for all permutations of values within blocks (i.e., the P-value of δ_o is 1 without alignment). The median values for blocks 1 and 2 are 3 and 8, respectively. If the median value is subtracted from the values of each block, then the aligned data are $y_{111} = 1$, $y_{112} = -1$, $y_{113} = 0$, $y_{121} = 1$, $y_{122} = -1$, and $y_{123} = 0$. If $v = 1$, then $\delta_o = 0$ and statistic δ is 0, 2/3, and 4/3 with probabilities 1/6, 1/3, and 1/2, respectively (the P-value of δ_o is 1/6 with alignment). If y_{kij} represents an aligned set of data, then $x_{kij} = y_{kij}/\phi_k$ where

$$\phi_k = \left[\sum_{i_1=1}^{g} \sum_{i_2=1}^{g} \sum_{j_1<j_2} \left| y_{ki_1j_1} - y_{ki_2j_2} \right|^v \right]^{\frac{1}{v}} \tag{3.5}$$

for $k = 1, \ldots, r$. This commensurated data has the desired property that

$$\sum_{i_1=1}^{g} \sum_{i_2=1}^{g} \sum_{j_1<j_2} \left| x_{ki_1j_1} - x_{ki_2j_2} \right|^v = 1 \tag{3.6}$$

for $k = 1, \ldots, r$ and any choice of v. If $v = 1$, then this commensuration is Euclidean commensuration.

Since MRBP is a method for analyzing treatment differences in the presence of blocks and the majority of experimental designs are concerned with analyzing treatment differences in a blocking framework, MRBP provides a natural analysis method for the majority of experimental designs. If one or more values occur for each treatment-block combination, then MRBP can be applied by simply replacing the replicated values with the median value. An obvious added feature of MRBP is the analyses of experimental designs involving multivariate data.

3.1. *LAD regression prediction models*

As in Section 2.4, again consider the multiple linear regression model given by

$$y_i = \sum_{j=1}^{s} x_{ij}\beta_j + e_i \tag{3.7}$$

where y_i is the ith of N observed values, x_{ij} denotes the jth of s predictors associated with the ith observed value where $x_{i1} = 1$ if the model includes an intercept, β_j designates the jth of s regression parameters, and e_i represents the error associated with the ith observed value. Since the desired fit consistent with Euclidean geometry between the N predicted values $(\tilde{y}_1, \ldots, \tilde{y}_N)$ and the N observed values $(y_1, \ldots, y_N)$ minimizes

$$\sum_{i=1}^{N} |\tilde{y}_i - y_i|, \tag{3.8}$$

the estimators of $\beta_1, \ldots, \beta_s$ which yield

$$\tilde{y}_i = \sum_{j=1}^{s} x_{ij}\tilde{\beta}_j \tag{3.9}$$

for $i = 1, \ldots, N$ are the values $\tilde{\beta}_1, \ldots, \tilde{\beta}_s$ (see Section 2.4). However, the permutation test in this situation is a special case of MRBP involving $b = 2$ (i.e., two blocks corresponding to the observed and predicted values) and $g = N$ (i.e., the number of observed values). Here the MRBP statistic is

$$\delta = \frac{1}{N}\sum_{i=1}^{N} |\tilde{y}_i - y_i| \tag{3.10}$$

where the distribution of δ under H_0 assigns equal probabilities to each of the $M = (N!)^2$ possible allocations for the N positions associated with $\tilde{y}_1, \ldots, \tilde{y}_N$ and $y_1, \ldots, y_N$, i.e., $g = N$ and $b = 2$ in expression (3.3). Thus the δ statistic compares the within-group clumping of response measurements against the model specified by the random allocation under H_0. Since M is prohibitively large for even fairly small values of N, Pearson type III P-value approximations described in Section 2 are again utilized. Efficient computational expressions for obtaining μ, σ^2, and γ are contained in Mielke (1991), Mielke and Iyer (1982), and Mielke and Berry (1997b), the latter for the specific case when $b = 2$. Incidentally, expression (3.10) is termed mean absolute error (MAE) and is commonly used in atmospheric science (Elsner and Schmertmann, 1993; Mielke et al., 1996; Willmott, 1982; Willmott and Matsuura, 1995; Willmott et al., 1996).

In addition to a point estimate of a predicted value, an interval estimate may also be desired. Let

$$\tilde{y} = \sum_{j=1}^{s} x_j \tilde{\beta}_j \tag{3.11}$$

be the LAD regression predicted value of a univariate response variable y for a given set of s predictors, $x_1, \ldots, x_s$. Also, let $\tilde{e}_i = \tilde{y}_i - y_i$ $(i = 1, \ldots, N)$ denote N estimated univariate error values resulting from the LAD regression fit of N events. Then the empirical $1 - \alpha$ quantile interval for y is given by

$$\tilde{y} + \Phi_{\alpha/2} \leq y \leq \tilde{y} + \Phi_{1-\alpha/2} \tag{3.12}$$

where $0 \leq \alpha \leq 1$,

$$\Phi_p = \begin{cases} \tilde{e}_{1,N} & \text{if } p = 0, \\ \tilde{e}_{k,N} & \text{if } 0 < p < 1 \text{ and } k \neq Np, \\ 0.5(\tilde{e}_{k,N} + \tilde{e}_{k+1,N}) & \text{if } 0 < p < 1 \text{ and } k = Np, \\ \tilde{e}_{N,N} & \text{if } p = 1, \end{cases} \tag{3.13}$$

k is the positive integer closest to $Np + 0.5$ when $0 < p < 1$, and $\tilde{e}_{1,N} \leq \cdots \leq \tilde{e}_{N,N}$ are the N order statistics associated with $\tilde{e}_1, \ldots, \tilde{e}_N$.

Recent papers by Mielke et al. (1996, 1997) indicate substantial advantages for LAD regression over LSD (least squares) regression in meteorological applications. These papers explore the effect of sample size, various types of contamination, and the inclusion of uninformative predictors on the shrinkage of the agreement measure $\Re$ in expression (3.4), i.e., the reduction of the inflated agreement measure when data is fitted for a given sample to the anticipated agreement measure when the estimated parameters are used with other random samples of the same size. It is shown in Mielke et al. (1997) that a drop-one cross-validation estimator is a nearly unbiased estimator of shrinkage. Other investigators provide discussions of cross-validation in atmospheric science (Elsner and Schmertmann, 1994; Michaelsen, 1987; Willmott et al., 1996). This regression approach has been used to obtain forecast models of tropical cyclone activity (Mielke et al., 1996, 1997). Berry and Mielke (1998) compare LAD and LSD regression, document the robustness of LAD regression, contrast the intuitive nature of LAD regression with the counterintuitive nature of LSD regression, and examine the effects of distance, leverage and influence on LAD and LSD regression.

3.2. *Multivariate regression prediction models*

The regression models described in Subsection 3.1 were confined to univariate multiple linear regression models (i.e., regression models involving a single observed value for each of N cases and a linear function of the parameters associated with s predictors for each observed value). An extension consistent with Euclidean geometry to general multivariate regression models follows (i.e., $r \geq 2$ observed values for each of N cases and either a linear or nonlinear function of parameters associated with the s predictors for each of the r observed values).

Consider the general multivariate regression model given by

$$\mathbf{y}_i = \mathbf{h}(\boldsymbol{\beta}, \mathbf{x}_i) + \mathbf{e}_i \tag{3.14}$$

where $\mathbf{y}_i' = (y_{1i}, \ldots, y_{ri})$ denotes the row vector of r observed values for the ith of N events, $\mathbf{x}_i' = (x_{1i}, \ldots, x_{si})$ is the row vector of s predictor values for the ith event, $\boldsymbol{\beta}' = (\beta_1, \ldots, \beta_t)$ is the row vector of t parameters, $\mathbf{h}' = (h_1, \ldots, h_r)$ is the row vector of r model functions of $\boldsymbol{\beta}$ and $\mathbf{x}_i$ for the ith event, and $\mathbf{e}_i' = (e_{1i}, \ldots, e_{ri})$ denotes the r errors between the observed values and model functions for the ith event ($i = 1, \ldots, N$). The special case of a linear multivariate model is given by

$$\mathbf{y}_i = \mathbf{B}\mathbf{f}(\mathbf{x}_i) + \mathbf{e}_i \tag{3.15}$$

where $\mathbf{f}'(\mathbf{x}_i) = (f_1(\mathbf{x}_i), \ldots, f_p(\mathbf{x}_i))$ denotes the row vector of p distinct functions of the s predictors ($\mathbf{x}_i$) for the ith event ($i = 1, \ldots, N$) and $\mathbf{B}$ is an $r \times p$ matrix of parameters in which $(B_{j1}, \ldots, B_{jp})$ is the row vector of p parameters associated with the jth observed value ($j = 1, \ldots, r$). Note that the multivariate linear regression model in expression (3.15) is a linear function of p parameters for each of the r observed values.

If $\tilde{\mathbf{y}}_i' = (\tilde{y}_{1i}, \ldots, \tilde{y}_{ri})$ denotes the row vector of r predicted values for the ith of N events, the distance function between $\tilde{\mathbf{y}}_i$ and $\mathbf{y}_i$ underlying a multivariate regression procedure is considered. Let $\Delta(\tilde{\mathbf{y}}_i, \mathbf{y}_i)$ denote the distance function between the observed and predicted values and consider the broad class of distance functions given by

$$\Delta(\tilde{\mathbf{y}}_i, \mathbf{y}_i) = \left(\sum_{j=1}^{r} |\tilde{y}_{ji} - y_{ji}|^w \right)^{\frac{v}{w}} \tag{3.16}$$

where $w \geq 1$ and $v > 0$. Since $v = 1$ yields the Minkowski metric (Arabie, 1991), the choice of $v = 1$ is preferred since $v > 1$ yields a distance function which does not satisfy the triangle inequality property of a metric. The analysis space associated with the Minkowski metric is termed ℓ_w^r (Simmons, 1963). If $r \geq 2$, the desired property that analyses based on the Minkowski metric are invariant to coordinate rotation holds only when $w = 2$ (Mielke, 1987). Thus ℓ_2^r, which is also termed r-dimensional Euclidean space, is the preferred analysis space. Consequently, the distance function of choice utilizes $v = 1$ and $w = 2$, i.e., Euclidean distance.

If $v = 1$ and $w = 2$, then the general and linear column vectors of r predicted values given by

$$\tilde{\mathbf{y}}_i = \mathbf{h}(\tilde{\boldsymbol{\beta}}, \mathbf{x}_i) \tag{3.17}$$

and

$$\tilde{\mathbf{y}}_i = \tilde{\mathbf{B}}\mathbf{f}(\mathbf{x}_i), \tag{3.18}$$

respectively, which minimize

$$\sum_{i=1}^{N} \Delta(\tilde{\mathbf{y}}_i, \mathbf{y}_i) \tag{3.19}$$

are the multivariate LAD regression predicted values and $\tilde{\beta}$ and $\tilde{\mathbf{B}}$ are the respective multivariate LAD regression estimators of β and $\mathbf{B}$. As previously indicated, this version of multivariate LAD regression is consistent with Euclidean geometry. Also the permutation test based on the MRBP statistic is

$$\delta = \frac{1}{N}\sum_{i=1}^{N} \Delta(\tilde{\mathbf{y}}_i, \mathbf{y}_i) \tag{3.20}$$

and, other than MRBP now being in a multivariate rather than a univariate context, the corresponding discussion regarding the permutation test in Section 3.1 holds here. The use of search routines to maximize the measure of agreement $\Re$ in expression (3.4) provides an excellent alternative criterion to LAD regression (Mielke and Berry, 1997b), especially for complex theoretically-derived nonlinear multivariate regression models which arise in meteorology (Cotton et al., 1994; Kelly et al., 1989; Lee et al., 1995; Tucker et al., 1989).

4. Summary

Although statistical analyses were originally based on between-value Euclidean distances, the approach changed around 1800 with the advent of complex regression analyses. Subsequently, the use of squared Euclidean distances became dominant in contemporary statistical analyses. In addition, while permutation tests have enjoyed a long history in statistics, only recently have modern computing methods made possible their extensive use. Permutation methods lead naturally to the use of Euclidean distances, rather than squared Euclidean distances, among data points. In this paper, permutation-based inference methods consistent with Euclidean geometry are described and applied to atmospheric science research. The comparisons of inference techniques based on Euclidean and squared Euclidean distances presented in Section 2.1 are most disturbing. A single value can overwhelm analyses based on squared Euclidean distances, regardless of the sample size. This geometric property characterizes an inferential deficiency underlying most of the statistical methods in use today.

Two permutation techniques, MRPP and MRBP, are described. MRPP is applicable to commonly-termed completely randomized experimental designs where the individual objects of interest are randomly allocated, under the null hypothesis, to g treatment groups. MRPP provides an omnibus test of differences among the g treatments. Because MRPP is a permutation test, an exact P-value of test statistic δ being as or more extreme than the observed δ may be obtained under the null hypothesis. When the number of permutations becomes large, it is necessary to approximate the exact distribution of δ by a continuous distribution. In such cases, a Pearson type III distribution is used to obtain an approximate P-value under the null hypothesis. In addition to the P-value, a chance-corrected measure of agreement, $\Re$, among objects within groups is described. Because the permutation distribution of δ and $\Re$ are equivalent, the P-value for δ is also the P-value for $\Re$. Applications of MRPP to cyclic data, to autoregressive patterns, and to multivariate correlated data are described.

MRBP can be thought of as an extension of MRPP where the randomization of objects to treatments is accomplished within each of b specified blocks, under the null hypothesis. MRBP provides an omnibus test of differences among the g treatments with appropriate adjustments for blocking. The calculation of exact MRBP P-values is unreasonable when the number of permutations is large. As with MRPP, a Pearson type III distribution provides approximate P-values. A chance-corrected measure of agreement, $\Re$, esitmates the agreement between blocks for all treatments.

LAD regression minimizes the sum of the absolute Euclidean distances between observed and predicted values in either a univariate or a multivariate research situation. The use of LAD regression provides significant advantages over LSD regression whenever the response variable or at least one of the predictor variables contains extreme values. The advantages stem, in large part, from the Euclidean distance function and the corresponding Euclidean analysis space intrinsic to LAD regression models. The use of LAD regression, based on squared Euclidean distances, is not consistent with the desired versions of either MRPP or MRBP based on Euclidean distances. MRPP provides permutation tests for analyzing differences among groups of LAD regression residuals associated with treatments. This application encompasses a broad class of linear model experimental designs, including completely randomized, randomized blocks, factorial, split-plot, and Latin square designs. LAD regression utilizes MPBP for the purpose of evaluating prediction models with observed data in both a univariate and multivariate framework. The chance-corrected measure of agreement, $\Re$, is presented and evaluated as an index of the fit between observed and predicted values in LAD regression models. Because the $\Re$ agreement measure is a function of permutations of absolute deviations, it is entirely consistent with the Euclidean analysis space of LAD regression models and ideally suited as a measure of fit between the observed and predicted values.

References

Arabie, P. 1991. Was Euclid an unnecessarily sophisticated psychologist? Psychometrika, 56:567–587.

Barrodale, I. and Roberts, F.D.K. 1973. An improved algorithm for discrete ℓ_1 linear approximation. Society for Industrial and Applied Mathematics Journal on Numerical Analysis, 10:839–848.

Barrodale, I. and Roberts, F.D.K. 1974. Solution of an overdetermined system of equations in the ℓ_1 norm. Communications of the Association for Computing Machinery, 17:319–320.

Berry, K.J. and Mielke, P.W. 1983. Moment approximations as on alternative to the F test in analysis of variance. British Journal of Mathematical and Statistical Psychology, 36:202–206.

Berry, K.J. and Mielke, P.W. 1984. Computation of exact probability values for multi-response permutation procedures (MRPP). Communications in Statistics—Simulation and Computation, 13:417–432.

Berry, K.J. and Mielke, P.W. 1988. A generalization of Cohen's kappa agreement measure to interval measurement and multiple raters. Educational and Psychological Measurement, 48:921–933.

Berry, K.J. and Mielke, P.W. 1992. A family of multivariate measures of association for nominal independent variables. Educational and Psychological Measurement, 52:41–55.

Berry, K.J. and Mielke, P.W. 1998. Least sum of absolute deviations regression: Distance, leverage, and influence. Perceptual and Motor Skills, 86:1063–1070.

Boscovich, R.J. 1757. De litteraria expeditione per pontificiam ditionem, et synopsis amplioris operis, ac habentur plura ejus ex exemplaria etiam sensorum impressa. Bononiesi Scientiarum et Artum Instituto atque Academia Commentarii, 4:353–396.

Bowditch, N. 1809. Observations of the comet of 1807. Memoirs of the American Academy of Arts and Sciences, 3:1–17.

Brockwell, P.J. and Mielke, P.W. 1984. Asymptotic distributions of matched-pairs permutation statistics based on distance measures. The Australian Journal of Statistics, 26:30–38.

Brockwell, P.J., Mielke, P.W., and Robinson, J. 1982. On non-normal invariance principles for multi-response permutation procedures. The Australian Journal of Statistics, 24:33–41.

Brown, B.M. 1982. Cramer-von Mises distributions and permutation tests. Biometrika, 69:619–624.

Cotton, W.R., Thompson, G., and Mielke, P.W. 1994. Real-time mesoscale prediction on workstations. Bulletin of the American Meteorological Society, 75:349–362.

Denker, M. and Puri, M.L. 1988. Asymptotic behavior of multi-response permutation procedures. Advances in Applied Mathematics, 9:200–210.

Durbin, J. and Watson, G.S. 1950. Testing for serial correlation in least squares regression. Biometrika, 37:409–428.

Elsner, J.B. and Schmertmann, C.P. 1993. Improving extended-range seasonal predictions of intense Atlantic hurricane activity. Weather and Forecasting, 8:345–351.

Elsner, J.B. and Schmertmann, C.P. 1994. Assessing forecast skill through cross-validation. Weather and Forecasting, 9:619–624.

Fisher, R.A. 1935. The Design of Experiments. Edinburgh: Oliver & Boyd.

Gauss, C.F. 1809. Theoria Motus Corporum Coelestium: In Sectionibus Conicis Solem Ambientium. Hamburg: F. Perthes and I. H. Besser.

Gray, W.M., Landsea, C.W., Mielke, P.W., and Berry, K.J. 1992. Predicting Atlantic seasonal hurricane activity 6–11 months in advance. Weather and Forecasting, 7:440–455.

Hess, J.C. and Elsner, J.B. 1994. Extended-range hindcastes of tropical-origin Atlantic hurricane activity. Geophysical Research Letters, 21:365–368.

Kelley, T.L. 1935. An unbiased correlation ratio measure. Proceedings of the National Academy of Sciences, 21:554–559.

Kelly, F.P., Vonder Haar, T.H., and Mielke, P.W. 1989. Imagery randomized block analysis (IRBA) applied to the verification of cloud edge detectors. Journal of Atmospheric and Oceanic Technology, 6:671–679.

Laplace, P.S. 1789. Sur quelques points du système du monde. Mémoires de l'Académie Royale des Sciences de Paris: 1-87. Reprinted in Laplace, P. S. 1878–1912. Oeuvres Complètes de Laplace, 11:477–558. Paris: Gauthier-Villars.

Lee, T.J., Pielke, R.A., and Mielke, P.W. 1995. Modeling the clear-sky surface energy budget during FIFE 1987. Journal of Geophysical Research, 100:25585–25593.

Michaelsen, J. 1987. Cross-validation in statistical climate forecast models. Journal of Climate and Applied Meteorology, 26:1589–1600.

Mielke, H.W., Anderson, J.C., Berry, K.J., Mielke, P.W., Chaney, R.L., and Leech, M. 1983. Lead concentrations in inner-city soils as a factor in the child lead problem. American Journal of Public Health, 73:1366–1369.

Mielke, P.W. 1978. Clarification and appropriate inferences for Mantel and Valand's nonparametric multivariate analysis technique. Biometrics, 34:277–282.

Mielke, P.W. 1979. On asymptotic non-normality of null distributions of MRPP statistics. Communications in Statistics—Theory and Methods, 8:1541–1550. Errata: 10:1795; 11:847.

Mielke, P.W. 1984. Meteorological applications of permutation techniques based on distance functions. In Handbook of Statistics, 4, P.R. Krishnaiah and P.K. Sen (Eds.) Amsterdam: North-Holland, pp. 813–830.

Mielke, P.W. 1985. Geometric concerns pertaining to applications of statistical tests in the atmospheric sciences. Journal of the Atmospheric Sciences, 42:1209–1212.

Mielke, P.W. 1986. Non-metric statistical analyses: Some metric alternatives. Journal of Statistical Planning and Inference, 13:377–387.

Mielke, P.W. 1987. L_1, L_2 and L_∞ regression models: Is there a difference? Journal of Statistical Planning and Inference, 13:430.

Mielke, P.W. 1991. The application of multivariate permutation methods based on distance functions in the earth sciences. Earth-Science Reviews, 31:55–71.

Mielke, P.W. and Berry, K.J. 1982. An extended class of permutation techniques for matched pairs. Communications in Statistics—Theory and Methods, 11:1197–1207.

Mielke, P.W. and Berry, K.J. 1994. Permutation tests for common locations among samples with unequal variances. Journal of Educational and Behavioral Statistics, 19:217–236.

Mielke, P.W. and Berry, K.J. 1997a. Permutation covariate analyses of residuals based on Euclidean distance. Psychological Reports, 81:795–802.

Mielke, P.W. and Berry, K.J. 1997b. Permutation-based multivariate regression analysis: The case for least sum of absolute deviations regression. Annals of Operations Research, 74:259–268.

Mielke, P.W. and Berry, K.J. 1999. Multivariate tests for correlated data in completely randomized designs. Journal of Educational and Behavioral Statistics, 24:109–131.

Mielke, P.W., Berry, K.J., and Brier, G.W. 1981a. Applications of multi-response permutation procedures for examining seasonal changes in monthly sea-level pressure patterns. Monthly Weather Review, 109:120–126.

Mielke, P.W., Berry, K.J., and Johnson, E.S. 1976. Multi-response permutation procedures for a priori classifications. Communications in Statistics—Theory and Methods, 5:1409–1424.

Mielke, P.W., Berry, K.J., Landsea, C.W., and Gray, W.M. 1996. Artifical skill and validation in meteorological forecasting. Weather and Forecasting, 11:153–169.

Mielke, P.W., Berry, K.J., Landsea, C.W., and Gray, W.M. 1997. A single-sample estimate of shrinkage in meteorological forecasting. Weather and Forecasting, 12:847–858.

Mielke, P.W., Berry, K.J., and Medina, J.G. 1982. Climax I and II: Distortion resistant residual analyses. Journal of Applied Meteorology, 21:788–792.

Mielke, P.W., Brier, G.W., Grant, L.O., Mulvey, G.J., and Rosenzweig, P.N. 1981b. A statistical reanalysis of the replicated Climax I and II wintertime orographic cloud seeding experiments. Journal of Applied Meteorology, 20:643–659.

Mielke, P.W. and Iyer, H. 1982. Permutation techniques for analyzing multiresponse data from randomized block experiments. Communications in Statistics—Theory and Methods, 11:1427–1437.

O'Reilly, F.J. and Mielke, P.W. 1980. Asymptotic normality of MRPP statistics from invariance principles of U-statistics. Communications in Statistics—Theory and Methods, 9:629–637.

Portnoy, S. and Koenker, R. 1997. The Gaussian hare and the Laplacian tortoise: Computability of squared-error versus absolute-error estimators. Statistical Science, 12:279–300.

Robinson, J. 1983. Approximations to some test statistics for permutation tests in a completely randomized design. The Australian Journal of Statistics, 25:358–369.

Rudin, W. 1966. Real and Complex Analysis. New York: McGraw-Hill.

Sheynin, O.B. 1973. R.J. Boscovich's work on probability. Archive for History of Exact Sciences, 9:306–324.

Simmons, G.F. 1963. Introduction to Topology and Modern Analysis. New York: McGraw-Hill.

Smith, P.L., Johnson, L.R., Priegnitz, D.L., Boe, B.A. and Mielke, P.W. 1997. An exploratory analyis of crop-hail insurance data for evidence of cloud-seeding effects in North Dakota. Journal of Applied Meterology, 36:463–473.

Tucker, D.F., Mielke, P.W., and Reiter, E.R. 1989. The verification of numerical models with multivariate randomized block permutation procedures. Meteorology and Atmospheric Physics, 40:181–188.

Walker, D.D., Loftis, J.C., and Mielke, P.W. 1997. Permutation methods for determining the significance of spatial dependence. Mathematical Geology, 29:1011–1024.

Watterson, I.G. 1996. Nondimensional measures of climate model performance. International Journal of Climatology, 16:379–391.

Willmott, C.J. 1981. On the validation of models. Physical Geography, 2:184–194.

Willmott, C.J. 1982. Some comments on the evaluation of model performance. Bulletin of the American Meteorological Society, 63:1309–1313.

Willmott, C.J., Ackleson, S.G., Davis, R.E., Feddema, J.J., Kink, K.M., Legates, D.R., O'Donnell, J., and Rowe, C.M. 1985. Statistics for the evaluation and comparison of models. Journal of Geophysical Research, 90:8995–9005.

Willmott, C.J. and Matsuura, K. 1995. Smart interpolation of annually averaged error temperature in the United States. Journal of Applied Meteorology, 34:2577–2586.

Willmott, C.J., Robeson, S.M., and Janis, M.J. 1996. Comparison of approaches for estimating time-averaged precipitation using data from the USA. International Journal of Climatology, 16:1103–1115.

Wong, R.K.W., Chidambaram, N., and Mielke, P.W. 1983. Application of multi-response permutation procedures and median regression for covariate analyses of possible weather modification effects on hail responses. Atmosphere–Ocean, 21:1–13.

Paul W. Mielke, Jr. is a professor of Statistics at the Colorado State University. He received a Ph.D. degree from the University of Minnesota in 1963. His research interests include permutation techniques, contingency table analyses, weather modification verifications, and short term climatic predictions.

Kenneth J. Berry is a professor of Sociology at the Colorado State University. He received a Ph.D. degree from the University of Oregon in 1966. His research interests include the logic of inquiry, permutation statistical methods, and tropical cyclone forecasting.

Data Mining and Knowledge Discovery, 4, 29–41 (2000)

Bootstrapping to Assess and Improve Atmospheric Prediction Models

J. SUNIL RAO sunil@hal.epbi.cwru.edu
Department of Biostatistics, Case Western Reserve University

Editors: Timothy Brown and Paul Mielke, Jr.

Abstract. Bootstrapping is a simple technique typically used to assess accuracy of estimates of model parameters by using simple plug-in principles and replacing sometimes unwieldy theory by computer simulation. Common uses include variance estimation and confidence interval construction of model parameters. It also provides a way to estimate prediction accuracy of continuous and class-valued outcomes regression models. In this paper we will overview some of these applications of the bootstrap focusing on bootstrap estimates of prediction error, and also explore how the bootstrap can be used to improve prediction accuracy of unstable models like tree-structured classifiers through aggregation. The improvements can typically be attributed to variance reduction in the classical regression setting and more generally a smoothing of decision boundaries for the classification setting. These advancements have important implications in the way that atmospheric prediction models can be improved, and illustrations of this will be shown. For class-valued outcomes, an interesting graphic known as the CAT scan can be constructed to help understand the aggregated decision boundary. This will be illustrated using simulated data.

Keywords: bootstrap, CART, classification, hurricanes, instability, supervised learning, weather data

1. Introduction

We consider the (supervised) classification/regression problem. We are presented with a learning sample $\mathcal{L} = \{(y_i, \mathbf{x}_i), i = 1, \ldots, n\}$ generated from an unknown distribution F. The y_i are either class labels $y_i \in \{1, 2, \ldots, J\}$, or some continuous measure, and the $\mathbf{x}_i$ are p-vectors of input variables for the ith case (observation). The aim is to relate y to $\mathbf{x}$ through a general model

$$y = \psi(\mathbf{x}, \theta) + \epsilon \tag{1}$$

where the errors ϵ have mean zero, variance σ^2 and are independent across training cases. The ψ gives the function form of how the predictors $\mathbf{x}$ relate to y and is indexed through a parameter vector θ. A typical example would be the linear regression model where $\psi(\mathbf{x}, \theta) = \sum_j^p x_j \beta_j$, where the β_j's are slope parameters corresponding to each x_j. For future input $\mathbf{x}$, we would generate the predicted value $y(\mathbf{x}, \theta)$ based on (1). Since we don't know θ typically, we estimate it from $\mathcal{L}$ giving $\hat{\theta}$.

This is a very common setup for problems encountered in the atmospheric sciences. For example, consider the air pollution data taken from Chambers and Hastie (1991). The goal is to predict ozone concentration (in ppm) from solar radiation, temperature (degrees

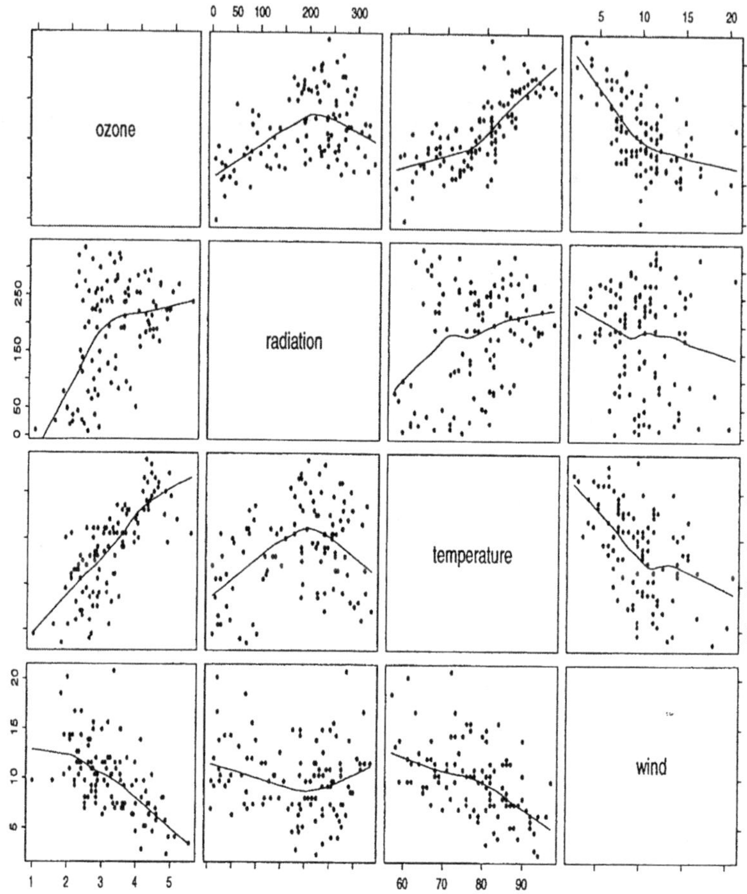

Figure 1. Pairwise scatterplots with loess smoothers for the predictors in the ozone data example. Variables are daily ozone concentration (ppm), solar radiation, temperature (degrees F) and windspeed (mph).

Fahrenheit) and wind speed (in mph) measured on 111 consecutive days in New York City. A scatterplot matrix for this data is presented in figure 1. A local regression non-parametric smoother (Cleveland, 1979) known as loess is overlayed on each scatterplot to give some indication of trends. It is quite clear for instance that ozone levels seem to rise linearly with temperature and decrease linearly with wind speed. Thus, one model for consideration may be,

$$ozone = \beta_0 + \beta_1\, rad + \beta_2\, temp + \beta_3\, wind + \epsilon \tag{2}$$

where the β' vector represents the regression coefficients and ϵ the random Gaussian error term.

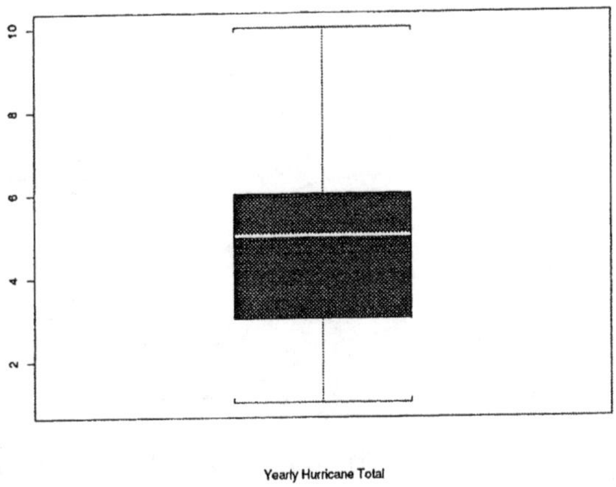

Figure 2. Boxplot of number of hurricanes over the 41 years of collected data from 1950–1990. The upper 25th percentile represents those seasons with greater than or equal to 6 hurricanes.

An example of a classification model would be the following. Hurricane prediction models are of great significance due to the wide-spread potential devastation caused by such storms. A dataset analyzed by Gray et al. (1992) contains hurricane data for the 41 year period 1950–1990. They tried to relate the number of hurricanes in a year (H) to the following predictors: Fiftymb = QBO wind component at 50mb (about 20 km above the surface in the tropics), Thirtymb = QBO wind component at 30mb (about 23 km above the surface in the tropics), RainG = average Gulf of Guinea African rainfall amount, and RainS = average Sahel African rainfall amount. Figure 2 shows a boxplot of the yearly count of hurricanes over the 40 years of collected data from the years 1950–1990. The upper 25th percentile corresponds to those years with hurricane counts of greater than or equal to 6 hurricanes and could potentially be considered heavy hurricane seasons (a dichotomous outcome). One possibly interesting model to relate this outcome to the 4 predictors would be a classification tree (Breiman et al., 1984; Quinlan, 1996).

A classification tree model describes the conditional distribution of response y given $\mathbf{x}$ and consists of two main components: a tree $T(\mathcal{L})$ and the parameter vector θ which associates a parameter θ_j with the jth terminal node. Trees are generated by recursively partitioning the predictor space into subsets (intermediate nodes) where the distribution of y is successively more homogeneous until some stopping point is reached (i.e., a terminal node). A simple cost-balanced pruning algorithm helps to minimize effects of over-training on $\mathcal{L}$. If in the resulting tree $\mathbf{x}$ lies in the region corresponding to the jth terminal node, then $y \mid \mathbf{x}$ has distribution $f(y \mid \mathbf{x})$ where f represents a parametric family indexed by θ. If y takes on class-values only, then $T(\mathcal{L})$ is known as a classification tree.

For illustration, we have only used the first 20 observations (the first half) of the dataset. The final tree is presented in figure 3. The numbers under the terminal nodes are the probabilities of *not* experiencing a busy hurricane season and the numbers within the nodes are the predicted tree outcomes for observations within that node. There are some interesting features that arise. In particular it appears that the busy hurricane seasons are associated

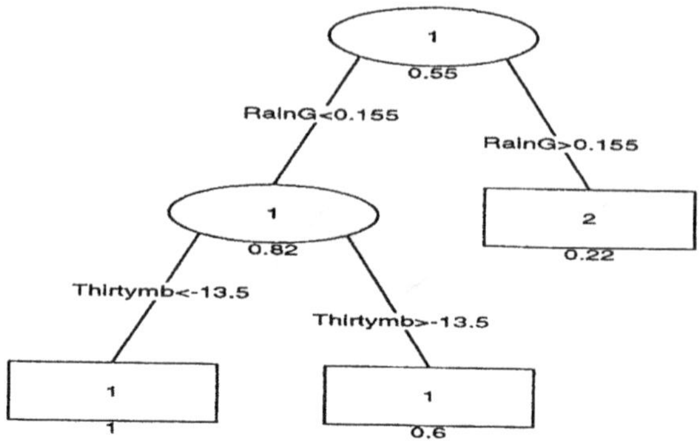

Figure 3. Classification tree for prediction of high hurricane seasons. Here a terminal node with predicted value of 1 in the node indicates that an observation landing in this node would be predicted to have greater than 6 hurricanes that year.

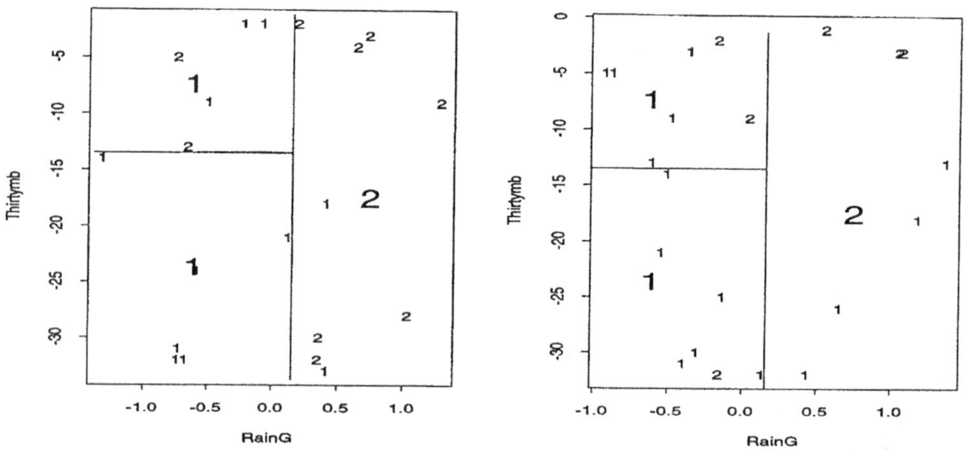

Figure 4. A two-dimensional partitioning of the covariate space based on the tree model with the predicted values for each partition from the tree in enlarged font size, 1 = not a busy hurricane season and 2 = a busy hurricane season. On the left is the training dataset of 20 observations and on the right is the test dataset also of 20 observations.

with RainG readings less than 0.155 and Thirty mb readings less than −13.5. In this case based on the small training dataset, there appears to be a .40 probability of experiencing a busy hurricane season. However for RainG readings of greater than .155, the probability of having a busy season jumps to .78. Figure 4 shows the partitioning of the covariate space based on the tree for both the training (left plot) and the test datasets (right plot). The plots show how tree-based models partition the covariate space along co-ordinate axes only.

Validation of these models is usually something of interest to an investigator and ideally a portion of the data would be set aside as a test dataset to be used for validating any models built on the training set. Prediction accuracy can then be assessed for example by looking at

$$PE = E_F(y - y(\mathbf{x}, \hat{\theta}))^2. \tag{3}$$

This is just the squared error loss function and for linear regression looks like average squared error. For the two-class outcome problem, it represents misclassification rate. When one is unable to efficiently set aside a test dataset, resorting to resampling techniques like the bootstrap (Efron, 1979) can prove to be very useful. These techniques amount to sampling repeatedly from the data at hand to generate multiple datasets with which to build and assess models with.

The rest of this paper will describe the bootstrap paradigm in general terms, prediction error assessment using the bootstrap and also some newer uses of the bootstrap in order to improve predictions by ensemble methods or aggregation over multiple candidate models. This proves particularly useful in situations where topological instability is a problem as in the classification tree example. Some graphical techniques for exploring these models will be shown that will more clearly demonstrate topological instability, and an approximate link to Bayesian model averaging will be discussed. It is also often the case in atmospheric sciences that one might want to use boostrap resampling in a more exploratory sense. This could include things like sensitivity analysis for parameters of a model and examining stability of models to deviations of the original dataset. The latter aspect will be examined in some detail in this paper. For more information on sensitivity analysis and other exploratory uses of the bootstrap, the reader can consult Leger et al. (1993) and Davison and Hinkley (1997).

2. The bootstrap paradigm

The bootstrap (Efron, 1979) is a computationally intensive technique for assessing among other things standard errors, biases and confidence intervals in non-standard situations. It replaces computational power for sometimes awkward statistical theory. The bootstrap paradigm is easily stated. A set of data $\mathcal{X} = (x_1, x_2, \ldots, x_n)$ is assumed to come from some unknown distribution F. A statistic can be defined as $S(\mathcal{X}, F)$. In order to do inference, the underlying sampling distribution of $S(\mathcal{X}, F)$ must be known. The bootstrap estimates F by some estimate $\hat{F}$ based on $\mathcal{X}$, and thus gives a sampling distribution based on $S(\mathcal{X}^*, \hat{F})$ where $\mathcal{X}^*$ is generated from $\hat{F}$. This is usually done via Monte Carlo simulation by repeated sampling from $\hat{F}$.

The bootstrap can be applied parametrically by assuming some form for F and estimating unknown parameters by their sample quantities. It can also be done non-parametrically where F is estimated by its non-parametric maximum likelihood estimate, the empirical distribution function which puts equal mass $1/n$ at each x_i. Sampling from $\hat{F}$ now means sampling with replacement from $\mathcal{X}$. Clever applications of this paradigm have been developed for many settings including linear and generalized linear models and neural networks (Shao and Tu, 1995; Tibshirani, 1997). Comprehensive overviews of bootstrap methodology can be found in Davison and Hinkley (1997), Efron and Tibshirani (1993), Hall (1992), and Shao and Tu (1995).

3. Bootstrap estimates of prediction accuracy

Assuming model (1), there are two different ways of bootstrapping (Efron and Tibshirani, 1993; Wu, 1986). One can consider each training case as a sampling unit, and sample with replacement from the training sets to create a bootstrap sample. This is called bootstrapping pairs. Let $\hat{y} = y(\mathbf{x}, \hat{\theta})$. One can then also consider the predictors as fixed and treat the model residuals $y_i - \hat{y}_i$ as the sampling units, and create a bootstrap sample by adding residuals to the model fit $\hat{y}_i$. This is known as bootstrapping residuals. The two algorithms are outlined below.

Note that each method requires refitting the model B times. The bootstrap residual procedure is model-based and relies on the fact that the errors $y_i - \hat{y}_i$ are representative of the true model errors. If the model is either mis-specified or overfit, the bootstrap pairs approach is more robust. The bootstrapping pairs approach can result in a different set of predictor values for each bootstrap sample, and in some settings, this may not be appropriate. In some situations, the set of predictor values is chosen by design and we wish to condition on those values in our inference procedure.

Prediction error estimates can be constructed from these resampling procedures in a number of ways.

Bootstrapping pairs:

1. Generate B samples, each one of size n drawn with replacement from the n training observations. Denote the bth sample by $\{(\mathbf{x}_1^{*b}, y_1^{*b}), (\mathbf{x}_2^{*b}, y_2^{*b}), \ldots, (\mathbf{x}_n^{*b}, y_n^{*b})\}$.
2. For each bootstrap sample $b = 1, 2, \ldots, B$, find the best estimate of the regression parameter giving $\hat{\theta}^{*b}$.

Bootstrapping residuals:

1. Estimate $\hat{\theta}$ from the training sample and construct residuals $r_i = y_i - y(\mathbf{x}_i, \hat{\theta}), i = 1, 2, \ldots, n$.
2. Generate B samples, each one of size n drawn from replacement from $r_1, r_2, \ldots, r_n$. Denote the bth sample by $r_1^{*b}, r_2^{*b}, \ldots, r_n^{*b}$ and let $y_i^{*b} = y(\mathbf{x}_i, \hat{\theta}) + r_i^{*b}$.
3. For each bootstrap sample find the best estimate of θ again.

Either method gives an estimate of (3) as

$$\widehat{PE}^B = \frac{1}{B}\sum_{b=1}^{B}\frac{1}{n}\sum_{i=1}^{n}(y_i - y(\mathbf{x}_i, \hat{\theta}^{*b}))^2. \tag{4}$$

Efron (1982) showed that in fact (4) can lead to an underestimation of PE of the order n^{-1}. He gave a less biased form making use of the bootstrap estimate of optimism. This is an attempt at correcting the underestimation of PE by the apparent error rate given by $ae = n^{-1}\sum_{i=1}^{n}(y_i - y(\mathbf{x}_i, \hat{\theta}))^2$. The correction known as the optimism in ae is given by $\omega = E_F(PE - ae)$. The natural bootstrap estimate of this is $\hat{\omega}^B = E_*(n^{-1}\sum_{i=1}^n (y_i - y(\mathbf{x}_i, \hat{\theta}^*))^2 - ae^*)$, where the expectation is simply with respect to $\hat{F}$. This estimator of PE can then be written as

$$\widehat{PE}_\alpha{}^{Bopt} = ae + \hat{\omega}^B, \tag{5}$$

and has been shown to be almost unbiased (Shao, 1996).

Another alternative is to focus solely on points that are not resampled for the purpose of prediction. This would avoid the overlapping of training and test data points. Consider applying the bootstrap in the following way. Let $(\mathbf{x}^*_{(i)}, \mathbf{y}^*_{(i)})$ be a bootstrap sample from the empirical distribution on all but the ith data pair. That is

$$\hat{F}_{(i)} : Pr \frac{1}{n-1} \; on \; (\mathbf{x}_1, y_1), (\mathbf{x}_2, y_2), \ldots, (\mathbf{x}_{i-1}, y_{i-1}), (\mathbf{x}_{i+1}, y_{i+1}), \ldots, (\mathbf{x}_n, y_n)$$

Applying this procedure to each case i in turn leads to the leave-one-out bootstrap given by

$$\widehat{PE}^{(1)} = \frac{1}{n}\sum_{i=1}^{n} E_{\hat{F}_{(i)}} Q_i,$$

where Q is the loss function under consideration. The actual calculation of this quantity is straightforward and just requires book-keeping of the original bootstrap samples from $\hat{F}$. Let N_i^b be the number of times that $(\mathbf{x}_i, y_i)$ is included in the bth bootstrap sample and define

$$I_i^b = \begin{cases} 1 & \text{if } N_i^b = 0 \\ 0 & \text{if } N_i^b > 0 \end{cases}$$

Then

$$\widehat{PE}^{(1)} = \frac{1}{n}\sum_{i=1}^{n} \hat{E}_i$$

where $\hat{E}_i = \sum_b \frac{I_i^b Q_i^b}{\sum_b I_i^b}$. This agrees with the leave-one-out definition because a bootstrap sample that has $I_i^b = 1$ is the same as a bootstrap sample from $\hat{F}_{(i)}$. Efron (1983) showed that there is an upward bias in $\widehat{PE}^{(1)}$ and proposed the .632 estimator:

$$\widehat{PE}^{(.632)} = .368(ae) + .632(\widehat{PE}^{(1)}). \tag{6}$$

The coefficients $.368 = e^{-1}$ and .632 were suggested by an argument based on the fact that bootstrap samples are supported on approximately $.632n$ of the original data points.

Some recent improvements to the accuracy of prediction error estimates have been made. These include the .632+ method of Efron and Tibshirani (1997). This basically entails

Table 1. Prediction errors for the ozone data regression model with $B = 25$.

Method	Prediction error estimate
Apparent error (ae)	0.251
Apparent error + Bootstrap optimism estimate	$0.251 + 0.020 = 0.271$
.632 Bootstrap	0.277
Leave-d-out cross-validation	0.268

calculating a weighted average as above with the weights now coming from a relative overfitting rate which puts greater weight on $\widehat{PE}^{(1)}$ in situations where the amount of overfitting as measured by $\widehat{PE}^{(1)} - ae$ is larger.

Example: The ozone data. Table 1 gives some PE estimates for the ozone data regression example. The apparent error (ae), bootstrap estimate of optimism correction, $\widehat{PE}^{(.632)}$ are presented. For comparison purposes, a leave-d-out cross-validation is presented where about $1/3$ (i.e. $d = .33$) of the sample is left out randomly to save as a test set for validation purposes (Efron and Tibshirani, 1993). The bootstrap analyses use only $B = 25$ bootstrap samples but still show some interesting facts. The apparent error (ae) clearly seems to be optimistic giving PE estimates approximately 8% lower than the .632 bootstrap and optimism corrections. This is clearly due to using all the same data for building and validating the model. The other less-biased estimates are reasonably close to one another in magnitude.

Many prediction models can be unstable to small perturbations of the training dataset. Examples of this include best subset regression and tree-based models like CART or C4.5. While this may be of some concern when trying to understand model topology, this instability can be capitalized upon. The next section outlines a new use of the bootstrap—to take advantage of this instability to (sometimes) greatly improve prediction accuracy of regression and classification models.

4. Bootstrap aggregation (bagging) for improved accuracy

Breiman (1996) introduced *bagging* (bootstrap aggregation) as a method to enhance the accuracy of unstable regression and classification methods like decision trees. Bagging takes advantage of instability by averaging over multiple trees to improve accuracy, but in the process destroys whatever interpretation the original method had (i.e. a bagged tree is no longer a tree). In bagging, bootstrap data sets, $\{\mathcal{L}^b, b = 1, \ldots, B\}$, are generated each consisting of n cases drawn at random, but with replacement from $\mathcal{L}$. A tree is grown on each dataset and the predicted value, $y(\mathbf{x}, \hat{\theta}^{*b}) = T(\mathbf{x}, \mathcal{L}^b)$ for a new input $\mathbf{x}$ (for example, from a test set $\mathcal{T}$, independent of $\mathcal{L}$), is obtained. A plurality vote among the B classifications is given by $T_B(\mathbf{x})$. Let $N_j = \#\{b : T(\mathbf{x}, \mathcal{L}^{(b)}) = j\}$ and then $T_B(\mathbf{x}) = argmax_j N_j$. Bagging decision trees have been shown to lead to generally consistent improvements in prediction accuracy for unstable classifiers (Breiman, 1996; Quinlan, 1996). Modifications have also been proposed where the resampling weights are continually updated according to the degree of difficulty of classifying training points. Those points that are harder to classify correctly get more weight and hence have a greater chance of appearing more frequently in bootstrap samples. This technique is known as boosting (Breiman, 1998; Freund and Schapire, 1996). So successful is this method that it has been called the best off-the-shelf classifier in the world.

Why does bagging work? To better understand this let the aggregated tree-predicted value at a new sample point $\mathbf{x}$ from F (i.e. averaged over repeated samples taken from F) be

$$T_A(\mathbf{x}) = E_F T(\mathbf{x}, \mathcal{L}) \tag{7}$$

Then average prediction error PE for a tree T is

$$PE = E_F(Y - T(\mathbf{x}, \mathcal{L}))^2 \tag{8}$$

and for the aggregated tree

$$PE_A = E_F(Y - T_A(\mathbf{x}))^2 \tag{9}$$

This type of depiction of prediction error is typical in the regression context where sum of squares of errors is typically used to assess model quality. It also has a natural generalization to the two-class classification problem (Tibshirani, 1996). For more general classification problems, more complex identities can be derived (Friedman, 1997). Focusing on the continuous response case, and using Jensen's inequality ($(EZ)^2 \leq EZ^2$) for some random variable Z, we can write,

$$PE \geq E(Y - T_A(\mathbf{x}))^2 = PE_A. \tag{10}$$

The more highly variable $T(\mathbf{x}, \mathcal{L})$ are, the more improvement aggregation may produce. The bagged estimate is just the bootstrap version of the aggregated one from F. If $\hat{F}$ approximates F well, then bagging will reduce PE's as well. The issue is not so clear in the general k-class problem where there is actually a trade-off of (boundary) bias and decreased variation that can determine bagging's potential gains (Friedman, 1997).

Example: The hurricane prediction classification tree revisited. The classification tree presented in figure 1 has a test set error rate of 35%. We bagged this training dataset of 20 observations and grew $B = 20$ trees on the bootstrap samples. Then by applying a plurality function vote for each test set observation, we were able to get the test set error rate down to 33%. While this represents only a relative reduction of 6% to the unbagged test set error rate, others have found much greater reductions in prediction error in various situations (Breiman, 1996). Looking at the right plot in figure 4, it is easy to see why in this situation, bagging does not provide large gains. The test set observations (randomly selected), are relatively well separated and it is just the cluster of 1's in the bottom right hand corner that causes some problems. So generally, co-ordinate axis splits of a single tree seem to do reasonably well in separating the two classes.As we will see in the subsequent section, bagging essentially smooths decision boundaries to make them more flexible and hence decrease prediction error.

4.1. *Understanding the bagged decision boundary: The CAT scan*

A useful understanding of bagging's microscopic properties can be elucidated using a new graphic called the classification aggregation tablet or CAT scan (Rao and Potts, to appear). This is a two-way array that focuses attention on the supervised learning problem and the decision boundary produced by bagging in the two-class problem. For each learning sample $\mathcal{L}$ and set of B bootstrap decision trees, a single CAT scan can be produced. It was developed to give a graphical explanation of why bagging works and was not intended as a data analysis tool (see figure 5).

The columns of the CAT scan are indexed by the number of bootstrap samples $b = 1, \ldots, B$. In this case, only $B = 50$ was used, and only the slices corresponding to $b = 1, 2, 5, 10, 50$ were displayed where by convention, $b = 1$ is the original dataset.

For reference, the true decision boundary (Bayes rule) and the learning sample are displayed in the title row. Each row (below the title row) of the CAT scan gives different types of displays of the decision boundary. The first row (i) gives the individual constituent classifiers corresponding to the bth sample that make up the aggregate. The decision regions

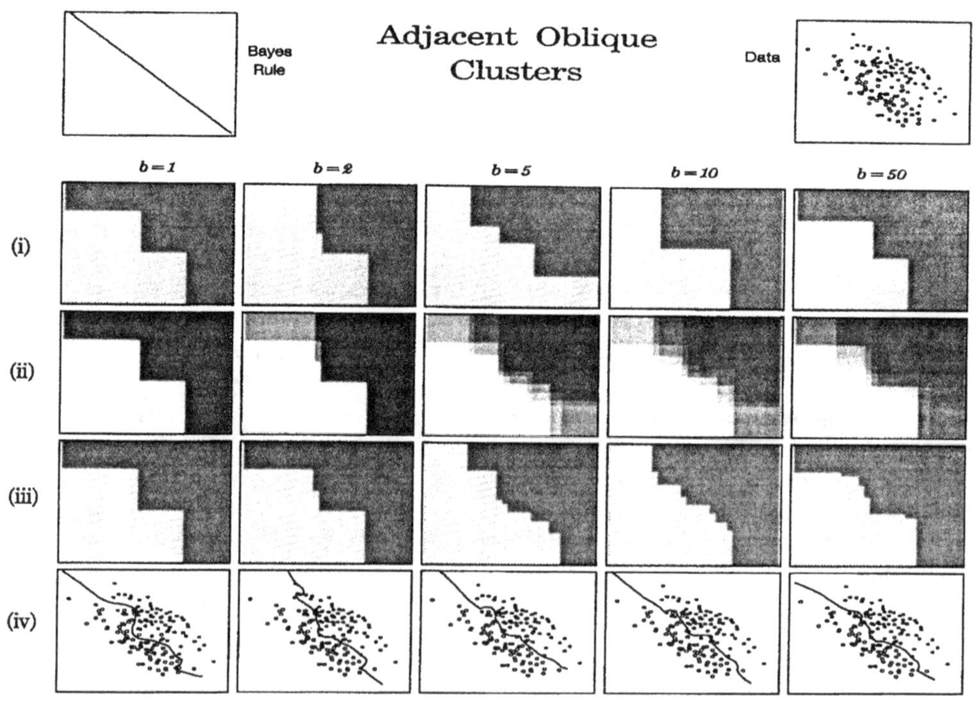

Figure 5. CAT scan for adjacent oblique clusters example as described in text.

are represented by grey shades, one for each class. The second row (ii) gives a cumulative aggregation of row (i). The number of grey shades equal the column index plus one, $b+1$. This represents the vote tally up to that column b. Consistent grey shades for each voting percentage were used across the row. They range from white, 100% of the votes one class, to black, 100% of the votes for the other class. The third row (iii) gives the cumulative, bagged decision boundary represented by two grey shades. The fourth row (iv) also represents the cumulative, bagged decision boundary as an interpolated contour plot. Further details on the construction of CAT scans can be found in Rao and Potts (to appear).

Example: Adjacent oblique clusters. Samples of size 75 for each of two classes were drawn from two bivariate normal populations. Let (μ_{11}, μ_{12}) and (μ_{21}, μ_{22}) be the mean vectors of input variables for the first and second class, respectively. Let $(\sigma_{11}, \sigma_{12})$ and $(\sigma_{21}, \sigma_{22})$ be the vector of standard deviations for first and second class, respectively. Let ρ_1 and ρ_2 be the correlations between the two input variables for the first and second class, respectively.

$$
\begin{aligned}
(\mu_{11}, \mu_{12}) &= (.4, .4), (\mu_{21}, \mu_{22}) = (.6, .6) \\
(\sigma_{11}, \sigma_{12}) &= (\sigma_{21}, \sigma_{22}) = (.2, .2) \\
\rho_1 &= \rho_2 = -.8
\end{aligned}
$$

In figure 5, the first row of the CAT scan shows some of the topological instability that exists with single tree models. The decision boundary varies to some degree from panel to panel corresponding to different looking trees (different in size and shape). The effects of bagging can also be seen. A smoothing of the decision boundary created by fine enhancements (small steps) along the boundary line are clearly evident. What is also clear from studying the simulated example, is that bagging is not a black box. It can be thought of a member of the general class of flexible discriminants (Ripley, 1996). It gives a flexible decision boundary with the ability to effectively model oblique and nonlinear Bayes rules.

4.2. *Bagging and Bayesian model averaging*

In this section we describe the connection between bagging and Bayesian model averaging which has enjoyed a renewed interest and copious amounts of recent research, primarily due to the Markov Chain Monte Carlo simulation methods for generating complexed high dimensional posterior distributions.

Suppose that the distribution of our data $\mathcal{L}$ depends on parameters θ. We have a loss function $L(\hat{y}, y)$ and our goal is to predict y with minimum expected loss. Then if θ has prior distribution $p(\theta)$, and our sampling density is $p(\mathcal{L} \mid \theta)$, we can write the posterior distribution of the prediction at $\mathbf{x}$ as

$$p(y \mid \mathbf{x}, \mathcal{L}) = \int p(y \mid \mathbf{x}, \mathcal{L}, \theta) p(\theta \mid \mathcal{L}) \, d\theta \tag{11}$$

where $p(\theta \mid \mathcal{L})$ is the posterior distribution of θ given the data.

We restrict attention to finite distributions on L categories, with $f_\ell = ((\mathbf{x}_i, y_i) \in$ category $\ell)$. The parameter θ is a functional of the f_ℓ, written as $\theta(\mathbf{f})$. Let $\hat{\mathbf{f}} = (\hat{f}_1, \ldots, f_L)$ be the observed proportions of n sample points falling in each category. Assuming a diffuse Dirichlet prior distribution for the f_ℓ, the posterior distribution $\hat{\mathbf{f}} \mid \mathbf{f}$ is a Dirichlet distribution, which is well approximated by the boot-strap distribution—a scaled multinomial distribution $(1/n) \cdot \text{Mult}(n, \hat{\mathbf{f}})$. We would like to estimate the posterior average

$$E(y \mid \mathbf{x}, \mathcal{L}) = \int E(y \mid \mathbf{x}, \mathcal{L}, \theta) p(\theta \mid \mathcal{L})\, d\theta \tag{12}$$

which is taken with respect to the underlying distribution F. The bootstrap simply plugs-in $\hat{F}$ for F, and approximates this expectation by Monte Carlo simulation. Thus the bagged prediction estimate can be thought of as an approximate posterior average.

Recently Rao and Tibshirani (1997) extended bagging to deal with implicit tuning parameters and proposed a way of generalizing it by averaging in two directions—over the bootstrap samples and over the tuning parameter distribution. This method was known as the out-of-bootstrap and was shown to be applicable not only for generating model-averaged predictions, but also for feature selection.

5. Discussion

There are some very exciting applications of the bootstrap and other resampling methods cropping up in the data mining field. This review paper has only briefly described two possible uses of the bootstrap for validating and possibly improving often unstable prediction models such as classification and regression trees. Since prediction models play such a vital role in the atmospheric sciences, there is much scope for the use and refinement of this methodology. While there has been a major push on improving prediction accuracy of models, what has been somewhat ignored is that interpretation of aggregated models is often sacrificed. This is still an open area of research. Another possibility to gain interpretation was developed by Chipman (1998) where he uses the bootstrap to generate a number of candidate models and then extracts out a topologically stable one.

References

Breiman, L. 1996. Bagging predictors. Machine Learning, 26:123–140.

Breiman, L. 1998. Arcing classifiers. Annals of Statistics, 26:801–849.

Breiman, L., Friedman, J., Olshen, R., and Stone, C. 1984. Classification and Regression Trees. New York: Chapman and Hall.

Chambers, J. and Hastie, T. 1991. Statistical Models in S. Pacific Grove, CA: Wadsworth Brooks Cole.

Chipman, H. 1998. Making sense of a forest of trees, Technical report, Department of Statistics and Actuarial Science, University of Waterloo.

Davison, A.C. and Hinkley, D.V. 1997. Bootstrap Methods and their Application. Cambridge, UK: Cambridge Univeristy Press.

Efron, B. 1979. Bootstrap methods: Another look at the jackknife, Annals of Statistics, 7:1–26.

Efron, B. 1982. The jackknife, the bootstrap, and other resampling plans. SIAM, Philadelphia.

Efron, B. and Tibshirani, R. 1993. An Introduction to the Bootstrap. New York: Chapman and Hall.

Efron, B. and Tibshirani, R. 1997. Improvements on cross-validation: the .632+ bootstrap method, Journal of the American Statistical Association, 92:548–560.

Freund, Y. and Schapire, R. 1996. Experiments with a new boosting algorithm, Machine Learning: Proceedings of the Thirteenth International Conference, pp. 148–156.

Friedman, J. 1997. On bias, variance, $0 - 1$ loss, and the curse of dimensionality, Journal of Data Mining and Knowledge Discovery, 1:55–77.

Gray, W.M., Landsea, C.W., Mielke, P.W., and Berry, K.J. 1992. Predicting Atlantic seasonal hurricane activity 6–11 months in advance, Weather and Forecasting, 7:440–455.

Leger, C., Politis, D.N., and Romano, J.P. 1993. Bootstrap technology and applications, Technometrics, 34:378–398.

Quinlan, R. 1996. Bagging, boosting, and C4.5. Proceedings of the AAAI National Conference on Artificial Intelligence.

Rao, J.S. and Potts, W. (to appear). Visualizing bagged decision trees, Journal of Computational and Graphical Statistics.

Rao, J.S. and Tibshirani, R. 1997. The out of bootstrap method for model averaging and selection, Technical report, Department of Statistics, University of Toronto.

Ripley, B. 1996. Pattern Recognition and Neural Networks. Cambridge, UK: Cambridge University Press.

Shao, J. 1996. Bootstrap model selection, Journal of the American Statistical Association, 91:655–665.

Shao, J. and Tu, D. 1995. The Jackknife and Bootstrap. Springer Series in Statistics, New York.

Tibshirani, R. 1996. Bias, variance and prediction error for classification rules, Technical report, Department of Statistics, University of Toronto.

Tibshirani, R. 1997. A comparison of some error estimates for neural network models, Neural Computation, 8:152–163.

Wu, C.J.F. 1986. Jackknife, bootstrap and other resampling plans in regression analysis (with discussion), Annals of Statistics, 14:1261–1350.

J. Sunil Rao is an assistant professor of Biostatistics at Case Western Reserve University. He received his Ph.D. from the University of Toronto in 1994 in Biostatistics. His research interests include theory and applications of bootstrap resampling, understanding ensemble models, model diagnostics for tree-based models, model selection and general biostatistics with particular attention to cardiovascular disease.

Data Mining and Knowledge Discovery, 4, 43–67 (2000)

Using Linked Micromap Plots to Characterize Omernik Ecoregions

DANIEL B. CARR dcarr@galaxy.gmu.edu
Center for Computational Statistics, George Mason University, Fairfax, VA 22030, USA

ANTHONY R. OLSEN tolsen@heart.cor.epa.gov
US Environmental Protection Agency, NHEERL Western Ecology Division, 200 S.W. 35th Street, Corvallis, OR 97333, USA

SUZANNE M. PIERSON spierson@mercury.cor.epa.gov
OAO, c/o US EPA, 200 S.W. 35th Street, Corvallis, OR 97333, USA

JEAN-YVES P. COURBOIS courbois@stat.orst.edu
Department of Statistics, Oregon State University, Corvallis, OR 97331, USA

Editors: Timothy Brown and Paul Mielke, Jr.

Abstract. The paper introduces linked micromap (LM) plots for presenting environmental summaries. The LM template includes parallel sequences of micromap, label, and statistical summary graphics panels with attention paid to perceptual grouping, sorting and linking of the summary components. The applications show LM plots for Omernik Level II Ecoregions. The summarized United States continental data includes USGS digital elevation, 30-year normal precipitation and temperature, and 8 million AVHRR pixels classified into 159 types of land cover. One LM plot uses a line-height glyph to represent all 159 land cover percentages per ecoregion. LM plots represent new visualization methodology that is useful in the data and knowledge based pattern representation and knowledge discovery process. The LM plots focus on providing an orienting overview. The overview provides a starting place for subsequent drilling down to what could otherwise be viewed as an overwhelming mass of data. The overview also provides a starting place to learn about the intellectual structure that lies behind the notion of ecoregions and begins to connect this abstract structure to quantitative methods.

Keywords: statistical graphics, LM plots, multivariate, spatial data, Omernik ecoregions, row plots, boxplots, line-height plots

1. Introduction

This visualization paper serves three narrow goals. First, we describe the template that underlies linked micromap (LM) plots. LM plots are a new way to represent spatially indexed statistical summaries. Second, we demonstrate the use of LM plots and other graphics for a complex task, characterizing continental United States ecoregions. The plots provide a glimpse of the scientific characterization of ecoregions, an ongoing process subject to revision and extension. Third, we call attention to a few selected modeling, summarization, and graphical issues that arise along the way.

A significant feature of the paper is the visualization of multiple large data sets composed of conceptually interrelated variables that are inherently spatial. We focus on providing a rich multivariate overview that is accessible to interested scientists. Scientists would use the overview to initiate their investigations. The overview may also serve as a portal to progressively more detail that can shed light on the conceptual models, mathematical models, and data collection methods. The graphics in this paper illustrate templates that are useful for overviews involving spatially indexed statistical summaries.

Providing a multivariate overview of something as complex as ecoregions must be tempered by limitations in human perception and cognition. Kosslyn (1994) states, "The spirit is willing but the mind is weak." The graphics here go far beyond Kosslyn's examples that typically involve twelve or fewer observations. While scientists are our target audience, they are very much human and scientific graphics are often subject to the criticism of attempting to show too much. With human limitations in mind, the current designs strive for apparent visual simplicity (Carr and Olsen, 1996). The general design task is not new to map designers (Bertin, 1983; Dent, 1990; MacEachren, 1994, 1995; Monmonier, 1991, 1993). Most design challenges are well known. For example, decisions must be made about which information to hide, how to encode information, and how to layer information. LM plots involve different tradeoffs than those encountered in cartography and constitute a new template (Carr et al., 1998). Previous presentations suggest that template interpretation requires modest learning and that after brief explanation, the graphics are accessible to scientists.

The task of providing a multivariate overview motivates thought about choice of representation media. We consider three alternatives: large hardcopy posters, web-based presentations, and journal articles. Previously, we used a 4×8 foot poster presented at the 1996 Joint Statistical Meeting in Chicago to introduce LM plots (Olsen et al., 1996). The poster included two large maps and many LM plots involving over 100 micromaps. The size, use of color, and number of graphics were not as constrained as in a journal. Journal articles are often constrained to examples that hint at what can be done.

Another alternative is to put the color graphics on the web. Web sites do present graphic problems. Tufte (http://www.clbooks.com/nbb/tufte.html) comments that, "the problem with the web is that it is low resolution in both space and time." Consider one map on the previously mentioned poster. This was an 8 million pixel color-coded land cover class map (159 classes) derived from AVHRR imagery. A typical workstation monitor can show $1024 * 1280$ pixels (roughly 1.3 million pixels). Viewing the image without reducing resolution requires panning over six complete screens, which is not ideal in terms of obtaining an overview. While the human computer interface community may be quick to point out that human pan and zoom does not provide logical progressive disclosure and other benefits of interactivity, much is to be said for large photographic quality posters as a suitable media for presenting visual overviews. We have created a site (www.galaxy.gmu.edu/~dcarr/ecoregions/) with the graphics for this paper including color alternatives for grayscale figures and additional examples referred to in the paper.

Scientific overviews can be works of science and art. Two recent book length overviews (Artic Monitoring, 1997 and Wahlstrom et al., 1996) involved the work of many people including layout specialists and graphics designers. The integration of photographs, text, and statistical graphics serves as an inspiration. Extended overviews that make frequent use of statistical graphics include Dorling (1995), Grant (1993), Pickle et al. (1997) and Zawitz (1983). This is not to endorse all the graphical designs, but much can be learned from these serious efforts to communicate a complex body of information and knowledge.

The world of interactive computer graphics also provides a rich source of overviews, often in the context of providing data base access and visual analysis tools. Eick (1996) presents many innovative single screen overviews for a variety of tasks. Greene et al. (1997a, b) and Plaisant et al. (1998) describe a few overviews developed in the human computer interface community. The graphics in this article are static, but comments in the article reflect current research directions that will move the LM plot into an interactive environment.

Having provided some of the general graphics context, we proceed in Section 2 to provide background on ecoregions and data sets. Section 3 contains an introduction to LM plots. Section 4 focuses on four LM plot examples and adds one glyph plot for contrast. Section 5 concludes by connecting LM plots to the research of others and by indicating challenges for the future.

2. Ecoregions and data sets

The graphics in this paper concern ecoregions. In this section we provide background on ecoregions and data sets to provide context for the subsequent graphics.

2.1. Ecoregions

Ecoregions are a way of codifying the recognizable regions within which we observe particular patterns or mosaics in ecosystems. General agreement is apparent that such ecological regions or ecoregions exist. However, disagreement in how the regions should be constructed continues to be a scientific issue (Omernik, 1995). In describing ecoregionalization in Canada, Wiken (1986) stated:

> Ecological land classification is a process of delineating and classifying ecologically distinctive areas of the earth's surface. Each area can be viewed as a discrete system which has resulted from the mesh and interplay of the geologic, landform, soil, vegetative, climatic, wildlife, water and human factors which may be present. The dominance of any one or a number of these factors varies with the given ecological land unit. This holistic approach to land classification can be applied incrementally on a scale-related basis from very site-specific ecosystems to very broad ecosystems.

Within the United States two alternative approaches in the construction of ecoregions are those developed by Bailey (1995a, b, 1998) and Omernik (1987, 1995). Each constructs a hierarchy of ecoregions that corresponds to viewing the United States at different scales. Omernik's approach is conceptually similar to that described by Wiken, where the ecological regions gain their identity through spatial differences in combinations of defining characteristics with the relative importance of each characteristic varying from one place to another and at all scales (different hierarchical levels). Bailey (1998) also has developed ecological regions hierarchically. First, he identifies ecological regions of continental scale based on macroclimate, where macroclimates influence soil formation, help shape surface topography, and affect the suitability of human habitation. The continent is subdivided with three levels of detail into domains, within-domain divisions, and within-division provinces. Domains and divisions are based largely on broad ecological climatic zones while provinces further sub-divide the divisions on the basis of macro features of the vegetation. Hence Bailey uses macroclimate as the controlling factor in the formation of ecoregions while Omernik uses all available factors where the importance of the factors varies among ecoregions. Some scientists question whether enough is known to delineate ecoregions. While knowledge is limited, some scientists proceed on the basis that their approximations of the "true" ecoregions will continue to improve as more information is gathered.

We have elected to use Omernik's level II ecoregions (figure 1a) for our work. Our interest is not to define ecoregions or even to validate them. We simply believe that it is important to describe quantitatively the key characteristics associated with ecoregions to gain a better understanding of how ecoregions have partitioned these characteristics. We do not have a comprehensive set of variables, but only a few variables that were readily available in digital form. Variables we use are land cover, 30-year normal annual precipitation, 30-year normal annual mean temperature, 30-year normal annual growing degrees days, 36 years of annual precipitation, and digital elevation (see web site examples).

2.2. *Data sets*

Loveland's Land-Cover. Loveland (1995) developed land cover regions for the United States from multi-date, meteorological satellite imagery acquired by the National Oceanic and Atmospheric Administration's (NOAA) Advanced Very High Resolution Radiometer (AVHRR) sensor. The data consist of 159 land cover classes for the approximately 8 million 1-km pixels, which cover the continental United States. The production of the land-cover data set began with the compilation of eight monthly maximum Normalized Difference Vegetation Index (NDVI) composite images covering the period March–October 1990 along with data on elevation, climate, and ecoregions. A clustering algorithm was applied to the eight NDVI images yielding 70 spectral-temporal classes. The additional data were used to help separate different types of vegetation cover that can not be distinguished from spectral characteristics alone.

Digital Elevation. Digital elevation data were obtained from the 1 : 250, 000-scale Digital Elevation Models (DEM), also known as 1-Degree DEM's, available from the U.S. Geological Survey (USGS, 1985).

30-Year Normal Climate. Nationally consistent climate data sets were obtained from PRISM (Parameter-elevation Regressions on Independent Slopes Model). PRISM, described by Daly et al. (1994), is an analytical model that uses point data and a DEM to generate gridded estimates of event, monthly and annual climatic parameters. PRISM models data

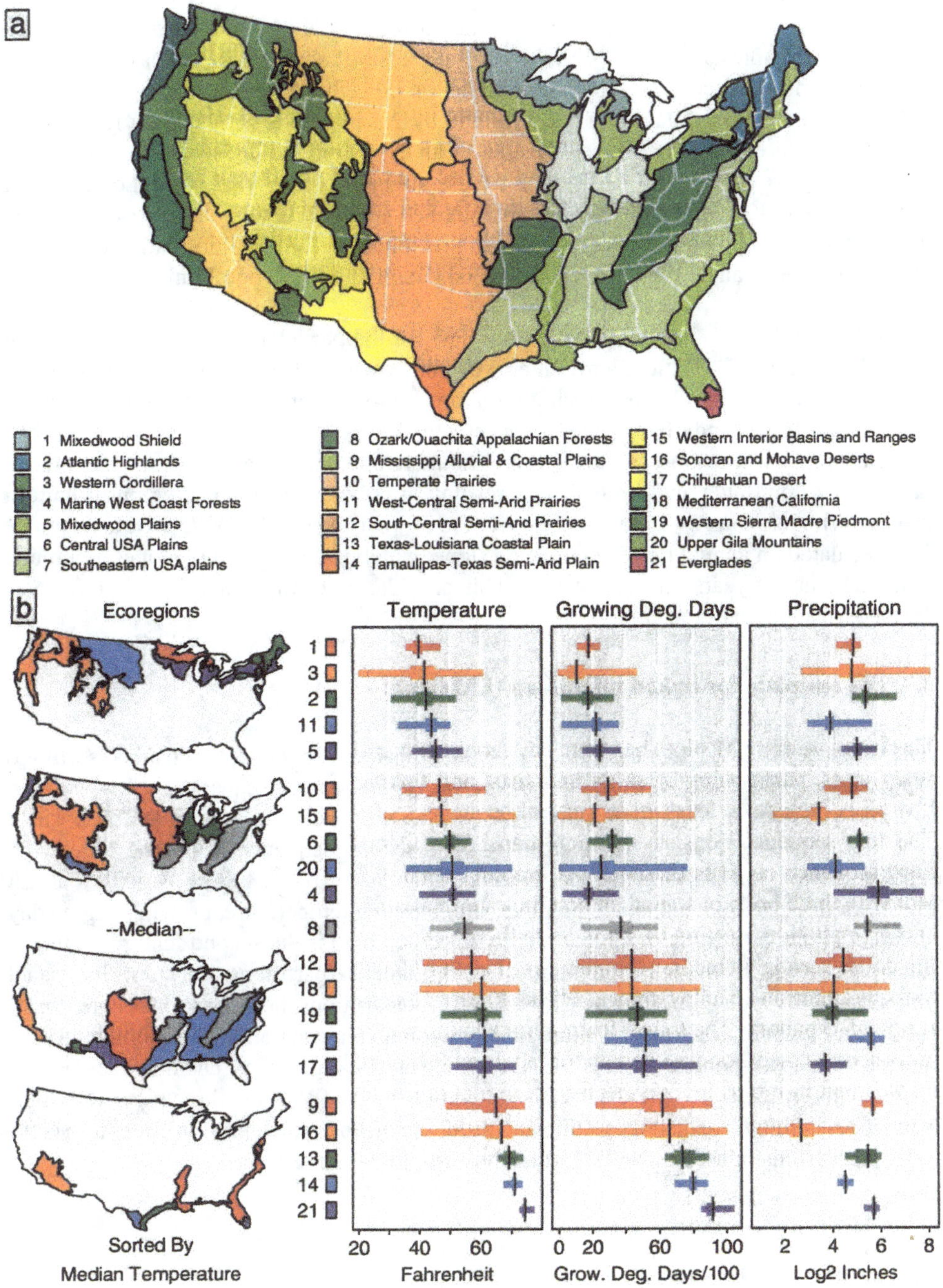

Figure 1. a = Omernik ecoregions, b = Linked micromap boxplots.

from individual climate stations to a regular grid through locally weighted climate/elevation regression functions. Local regions are defined using the DEM so that orographic effects on climate variables are included in the local regression models. PRISM has been shown to out-perform other common spatial interpolation procedures such as kriging (Daly et al., 1994). The data are available on a 2.5 minute by 2.5 minute grid. Having a regular grid of climate data ensures that the entire area of an ecoregion is represented in any statistical summary we construct. PRISM data we use are based on 30-year normal climate data for 1961–1990. Key annual parameters investigated are mean temperature, growing degree days, total annual precipitation, and monthly precipitation. Further information on PRISM and the data sets can be found at www.ocs.orst.edu/prism/prism_new.html.

Annual Precipitation. Annual precipitation data for the period 1961–1996 were obtained from the National Climatic Data Center web site, www.ncdc.noaa.gov/ol/climate/online/coop-precip.html#FILES. The precipitation data and weather station location descriptors involve 72 Mbytes and 10.2 Mbytes, respectively. We assigned weather stations to ecoregions using a point-in-polygon algorithm. The assignment used each station's most recent latitude and longitude and the 672 high-resolution polygons of the ecoregions. Omitting stations with less than 24 annual values over the 36 years resulted in 4075 weather stations. We calculated an annual trend using Sen's slope, which is the median of slopes, computed using all pairs of years for each station. This provides an estimate that is little influenced by an occasional extreme value.

3. The template for linked micromap (LM) plots

The template for LM plots has four key features: parallel sequences of panels, sorting of study units, partitioning of study unit lists, and linking across sequences. First, standard LM plots include at least three parallel sequences of panels that are linked by position. The first sequence consists of micromaps. The second sequence is linking labels. The third sequence consists of statistical graphics such as boxplots. Figure 1b shows an LM plot with three boxplot statistical graphics summarizing temperature, growing degree days and precipitation. Sorting of study units (here ecoregions) is the second feature. Figure 1b illustrates sorting by median temperature. The third feature partitions study units into panels to focus attention on a few units at a time. Figure 1b partitions the twenty-one ecoregions in a 5-5-1-5-5 pattern. The fourth feature links micromap representations of highlighted study units across corresponding panels for labels and statistical graphics. Distinct hues serve as a color link in figure 1b. We discuss these features below. With the three types of parallel sequences in mind, we discuss sorting and grouping of study units first, and end this section with discussions of across panel linking and LM plot labeling.

3.1. *Study units, sorting, and perceptual grouping*

The study units in LM plots have names, map locations, and summary statistics. Here the study units are ecoregions. LM plots are intended to be quite general (Carr et al., 1998). Other entities such as states, nations, monitoring sites, network elements, genes on a portion of a chromosome, and even parts of the human body can serve as study units that appear in micromaps.

LM plots sort the study units. Often, a plotted univariate summary statistic provides the basis for sorting. For example, ecoregion median temperature provides the basis for sorting ecoregions in figure 1b. Figure 2 uses the same sorting criterion, but the univariate median temperature is not a part of the display. There is uncertainty associated with statistical estimates used to sort study units. Statistical issues associated with ranking in the presence

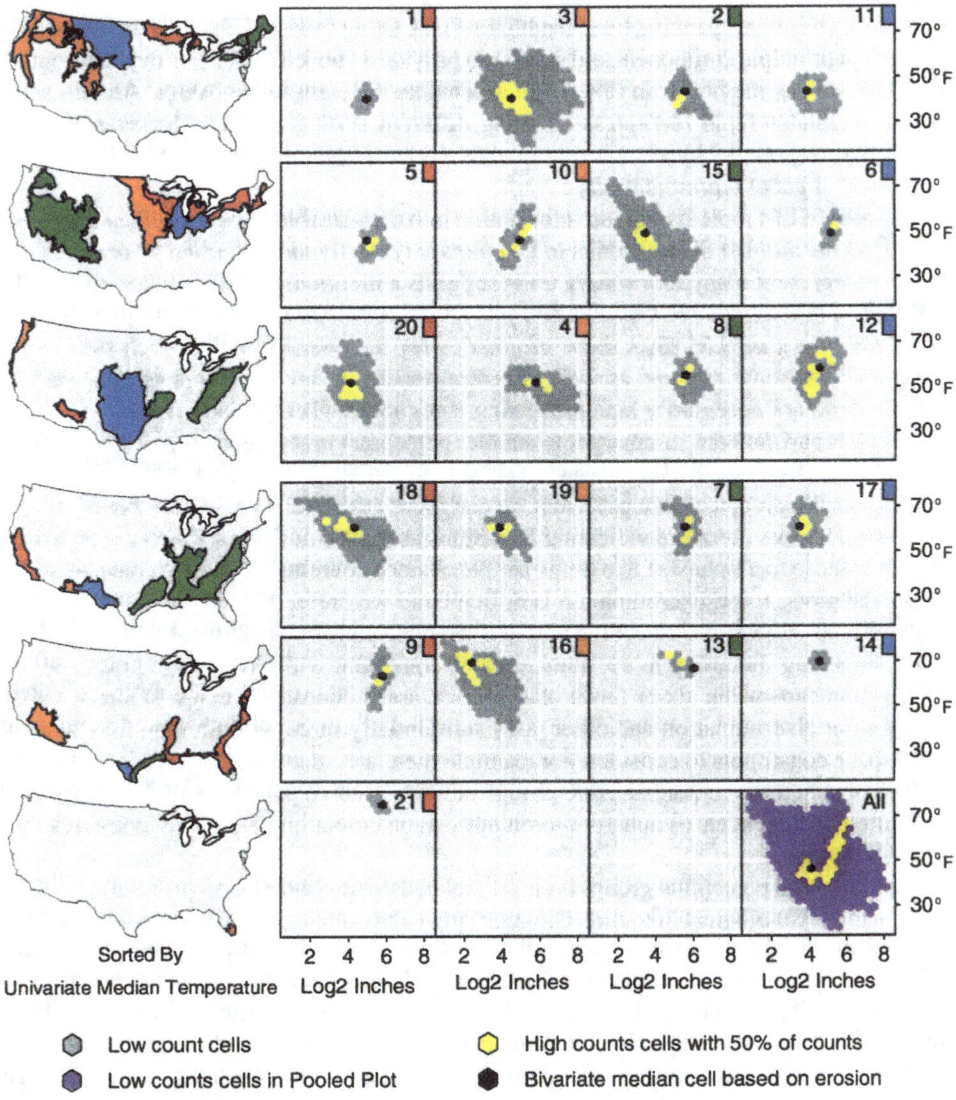

Figure 2. LM bivariate binned boxplots 1961–1990 precipitation (x) versus annual temperature (y).

of uncertainty are not fully appreciated although ranking difficulties are well known in sports. The confidence intervals shown in the LM dot plot of Carr and Pierson (1996) provides a visual indication of likely rank reversals. The ordering of study units in figures 1 and 2 should be considered as an approximation.

Many LM plots involve multivariate statistical summaries. For example, the statistical summaries show time series for monthly precipitation in figure 3 and area percents for 159 land cover classes in figure 4. Many multivariate sorting options are available for such cases. When the multivariate observations are in the same units, Cleveland (1993a) suggests sorting by the medians. This provided the sorting order for figure 3. Carr and Olsen (1996) promote ordering by minimal spanning tree breadth traversal. Other ordering methods include sorting by the first principal component and sorting to provide a convenient clustering dendrogram. Different sorting methods can reveal and accentuate different relationships. An interactive setting facilitates rapid re-expression using different sorting criteria. However, the easy ability to re-express LM plots does not obviate the need to think about the sorting order and the choice of panel representations.

One goal of LM plots is to group information into manageable units for human interpretation. Thus the number of study units in LM plots is typically modest, often 50 or fewer. The desire to represent many more study units suggests a hierarchical organization of multiple LM plots. For example one national LM plot might show summaries for all the U.S. states while multiple state LM plots show summaries for the counties within each state. Since overviews can show extrema as well as central tendency, say via boxplots with outliers, overviews do not necessarily hide interesting clues about where to look next.

LM plots partition the summary into smaller perceptual units. Carr (1994a) and Carr and Olsen (1996) suggest that long lists are visually intimidating. Kosslyn (1994) recommends creating small perceptual groups so that we can focus attention on a manageable number of elements. He cites literature indicating four items as a good number for direct comparison. We often chose to emphasize five items per panel since counting by fives is convenient. For gray scale plots, color discrimination is difficult and we prefer to keep the number to four. However, we stretch to five ecoregions in the center panels of figures 3 and 4. With full color, stretching the group to six units remains consistent with the cartographic literature that recommends using six or fewer classes for classed choropleth maps. While six items makes color discrimination and other tasks substantially more difficult than five there are often space constraints to consider. For example there are 78 level III Omernik ecoregions. Groups of six create 13 panels while groups of five create 16 panels. With fixed space, the size difference between 13 and 16 micromaps can be crucial in being able to see the color of small regions.

Creating small perceptual groups is closely related to the chunking of information that appears in the psychological literature. Creating small perceptual groups is also consistent with the human computer interface mantra of context and focus. In LM plots, the full sequence of panels provides one facet of context while an individual panel focuses attention on a few study units. Grouping of study units can be guided by different criteria. Carr et al. (1997), for example, grouped genes by gene function first and then created subgroups with four or fewer genes. Using logical criteria for sorting and grouping is often helpful. The cartographic practice of using gaps between values of the sorting variable to partition elements is

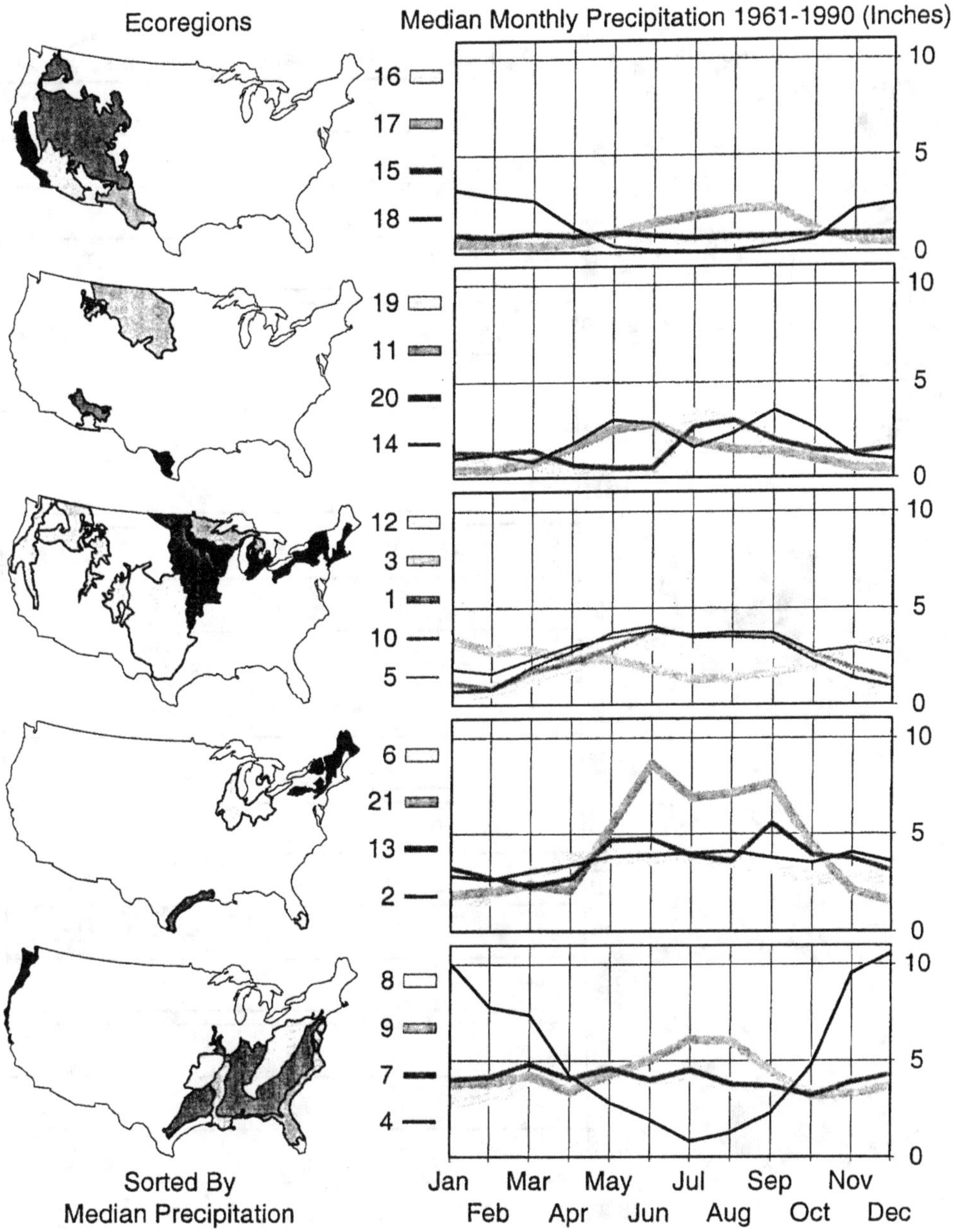

Figure 3. Linked micromap time series plot.

often advantageous for bringing out spatial patterns. Different objectives motivate different groupings.

Breaking a long list into smaller perceptual units can simplify visual appearance. The panel edges provide additional visual entry points. The reader may spot the name of a study

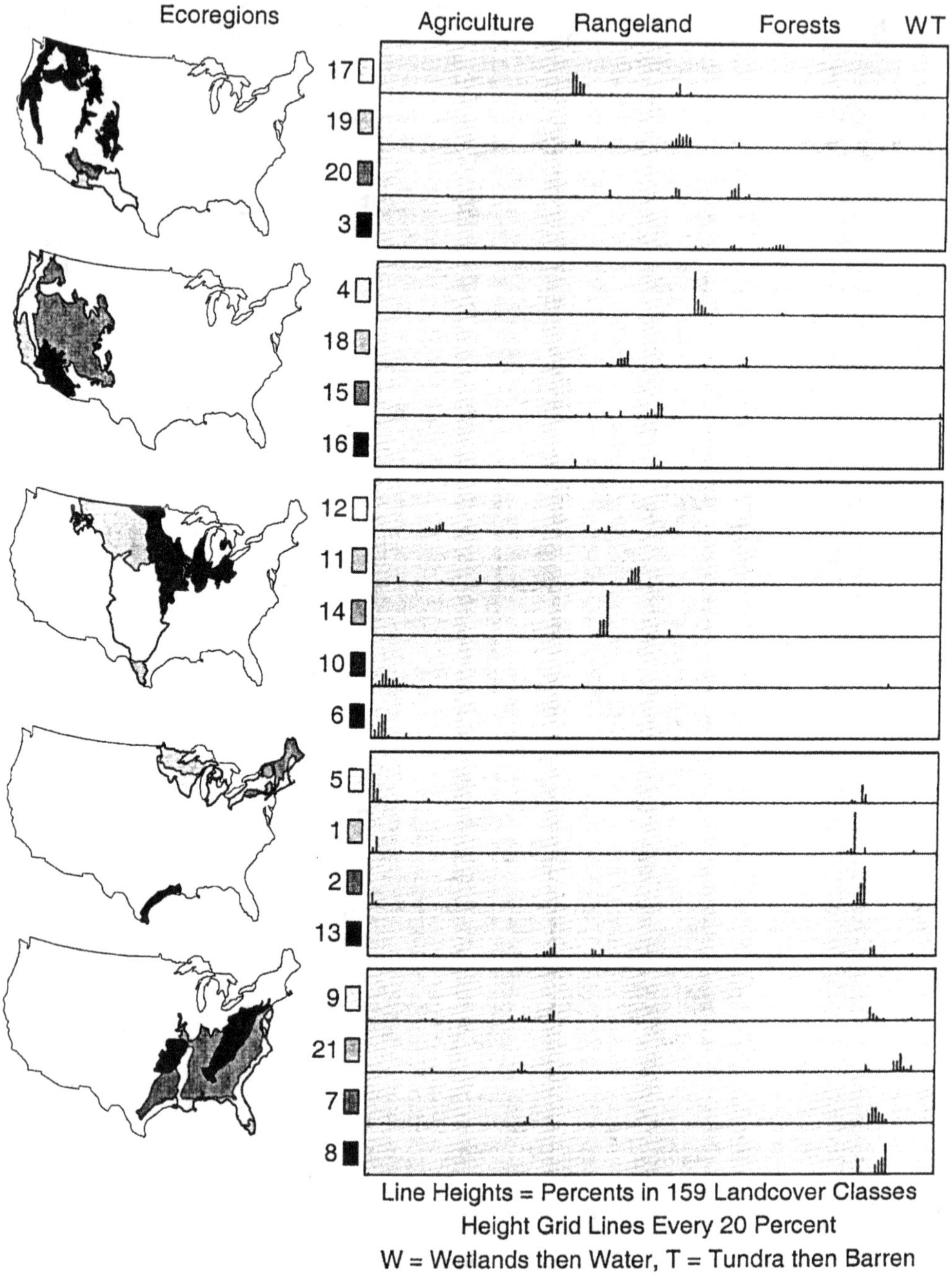

Figure 4. Linked micromap line height plot.

unit of particular interest at one of the panel edges and be drawn into the graphic. However, many small perceptual units can still constitute a visually intimidating list. Thus small perceptual units can be grouped into larger perceptual units. The design strategy of Carr and Pierson (1996) groups 50 states and Washington D.C. into one block with five groups of five, one block with one group of one and one block with five groups of five. The block in the middle contains the study unit that has the median value for the sorting variable. This study unit appears in both neighboring micromaps. More importantly the design splits the plot into two more manageable pieces. With twenty-one ecoregions, figure 1b uses a 5-5-1-5-5 pattern to get by with four micromaps. D. Andrew Carr developed a public domain Visual Basic front end to state LM dot plots and his defaults cover grouping through 51 items, but to our knowledge there is no general theory concerning iterated grouping for long lists.

3.2. Micromaps and linking

The primary task of micromaps is to show the general spatial location that corresponds to statistical estimates. Carr and Pierson (1996) indicate that micromaps are often spatial caricatures designed to serve specific purposes. The ecoregion micromaps are generalizations, i.e., simple caricatures, of the actual delineated ecoregions. The generalization includes simplified boundaries and reduces 682 polygons to 44 polygons. The simplified boundaries reduce unnecessary graphic clutter. In many situations, such as in the state micromaps of Carr and Pierson (1996), small study units are enlarged, while preserving neighbor relationships and shape detail for recognition, to facilitate perception of color.

Within a perceptual group, we use distinct saturated hues as a rapid link between statistical estimates, study unit names, and spatial locations. The cyclic use of distinct hues as a link to micromap regions often causes confusion when people first encounter LM plots. People expect the color encoding to remain the same for the entire sequence of micromaps. The local color linking of a name to a region in a micromap is a fast link since color is a texton handled in preattentive vision (Julesz, 1986). The easy link serves to re-enforce the knowledgeable reader's mental linking of study unit name to its spatial location. For some readers, the link serves as a reminder. For yet other readers, such as those encountering ecoregions for the first time, the link serves as an educational device. Thus there is a strong argument for using the local color link.

Our micromaps were first conceived in the context of providing a fast link to large maps with many study regions. After finding the shape and position of a region in a micromap it is very easy to find the region in a large map. Linking ecoregion name to location is difficult in figure 1a. There are many ecoregions and some involve disjoint areas. The colors are not easy to discriminate. For most readers, the colors do not have names, and this can make it much harder to remember the exact color that is to be found on a map. Plotting the region numbers on the map typically solves the memory problem, but serial search is still involved. With disjoint areas for the same ecoregion, finding one number does not mean the task has been completed. The problem gets worse with Level III ecoregions and number overplotting begins to be a problem. Micromaps were a good solution to the problem. However, we quickly noted that the patterns in the sequence of micromaps were informative. Since high-resolution boundaries were not our primary interest, we found the large maps to be of little value unless they contained additional spatial information, such as a representation of elevation.

The presence and absence of highly saturated hues reveals patterns in the sequence of micromaps. While the different hues convey ranking within a panel, the cyclic hues can be distracting when viewing the sequence of micromaps. In an interactive setting it is easy to modify the colors to help in revealing patterns. In figure 1b, the cool median temperatures are in the northern ecoregions as would be expected. The micromap sequence sorted by precipitation (available on our web site) shows a dramatic West to East pattern, with high median precipitation in the East. The single exception is the Marine West Coast Forests Ecoregion 4 that contains the Olympic rain forest.

Carr et al. (1998) augment micromaps with contours to bring out spatial patterns in a sequence of micromaps. In one state map variation they represent states above the median with light yellow fill and black outlines for panels above the median. This narrows the region of focus, provides stability across a subsequence of micromaps, and calls attention to a particular contour. The saturated color for highlighted states overrides the yellow and provides detail within the light yellow region. The remaining states are gray with white outlines as in figure 1b. The gray states appear as part of the background. (The web site precipitation-sorted example provides an illustration for a sequence of four micromaps like that in figure 1b.) The light yellow "spotlight" shifts to below median states for panels below the median. Logically the light yellow and gray regions are complements of each other in this variation. Nonetheless the light yellow region in the foreground gets much more attention. With some states at the map edges, switching the focus calls attention to patterns that were not noticed when the same composite region was in the background. The Carr et al. (1998) discussion of light yellow spotlights suggests additional variations that can help to bring out spatial patterns, especially in long sequences of micromaps.

More options for micromap color encoding remain to be investigated. Brewer (1997) discussed color scales for the color blind that are straightforward. After citing extensive literature arguing against spectral order, Brewer also cites perceptual studies indicating that spectral order works quite well. Spectral ordering (avoiding bright yellow) in combination with lightness ramps may yet provide useful color schemes for micromaps. We keep it simple here.

LM plots can supplement or replace classed choropleth maps so discussion is appropriate about what is lost and gained. Classed choropleth maps provide a quick gestalt impression. They can provide quick answers to certain questions, such as, "what are the values of a particular study unit and its neighbors?" Micromap sequences may not provide the immediate gestalt impression but they can show at least two contours very clearly while providing a complete ranking of study units. With LM plots it can take longer to find values for particular study units, but much more detail is available. The detail often includes multivariate descriptions. Note that choropleth maps often degrade continuous univariate statistical estimates by forcing them into a few classes and then represent the degraded estimates using a poor encoding, color. (See Cleveland and McGill, 1984 concerning perceptual accuracy of extraction and poor encodings.) The micromap approach trades off region boundary detail in exchange for more faithful representation of the statistical estimates that comprise the spatial surface. Wood (1992) discusses the agenda of maps. It is not surprising that cartographers and statisticians would have different agendas and choose different tradeoffs.

3.3. *Statistical summary panels*

The statistical summary panels can take many forms such as dotplots, barplots, boxplots, times series plots, scatterplots, density plots, cdf plots, perspective views, stereo pairs plots and so on. For example, figure 1b shows boxplots and figure 3 shows time series plots. While most of these plots are familiar, there are still many graphical design issues to address.

The small size of the statistics summary panels accentuates design challenges concerning perceptual accuracy of extraction, resolution, overplotting, and labeling. We discuss plot specific efforts to deal with design challenges in the examples sections below. A few design facets are common for all LM statistical summary panels, so are discussed here.

Cleveland and McGill (1984) indicate that position along a scale is the preferred graphical encoding in terms of perceptual accuracy. Thus all the LM examples use this encoding. Small panels lead to scales whose short length reduces the reader's ability to make fine distinctions. In the battle to promote perceptual accuracy of extraction we add background grid lines. The grid lines allow the reader to make judgments against a finer scale and thus increase the accuracy of judgments in accordance with Weber's law (Cleveland, 1994).

Our general preference is to use light gray background with white grid lines (Carr, 1994c). However, choice of other light background colors can be both effective and pleasing in terms of variety and color coordination. Occasionally we are pressed into using other colors, such as in figure 4 where gray level already encodes ecoregions.

Space constraints continually come into play. For memory and comparison reasons, it is often advantageous to keep a LM plot to a single page. We often forego Cleveland's (1993a) guidance about banking to 45 degrees and are tempted to skimp on labeling. Uncomfortable compromise, of course, is not unique to LM plots.

While there are many design issues, statistical summary panels can nonetheless accommodate much information that helps in interpretation. Carr and Pierson (1996) discuss the design of confidence bounds on dot plots. Carr et al. (1998) include reference values, for example using a dashed line to represent the U.S. average on a plot of state values. The designs can continue to evolve in ways that help the reader to make better use of the available information.

3.4. *LM plot labeling*

Labeling is one of the most difficult challenges in graphical design. Lack of explanation can mislead or confuse the reader. For example, it makes a difference if the underlying observations are measurements, simulated values, self-reported "guesstimates" as in EPA's Toxic Release Inventory, or outright fabrications. Good labeling answers questions about who, what, where, when, why, how and how well. Knowing the political and economic agenda behind data collection and reporting efforts can be more important than knowing the units of measure on a graph scale. Metadata inclusion in graphics is a problem that is easier to address in an interactive setting than in a static setting, (for example see Carr et al., 1996). In a static setting the metadata often appears in text that is easily separated from the graph. Carr (1994a) suggests developing icons to answer a small set of standard questions as a starting point. With limited space in LM plots we have done little with labeling. Our biggest change from early versions is inclusion of text indicating the study unit sorting order. For some, explanation of the color encoding would also be desirable. In terms of labeling much remains to be done.

4. Discussion of examples

We selected the five basic examples to illustrate different plot designs. Below we discuss facets of the designs not previously mentioned, along with discussion of plot interpretation.

4.1. Boxplots and figure 1b

The boxplot is a well-known distributional caricature that has now appeared in some grade school curricula. However, the choice of summary statistics used and their graphical representation are not universal (Carr, 1994b; Frigge et al., 1989; McGill et al., 1978; Tukey, 1993). Figure 1b shows a five-number summary: the extrema, the 1st and 3rd quartiles and the median. Another common choice represents adjacent values and outliers rather than extrema. We note that the outlier portion of this choice does not scale well for large samples from thick tailed distributions. For example if estimates followed a t-distribution with 3 degrees of freedom roughly 2.75 percent of the estimates would be flagged as outliers on each end of the distribution. Outlier overplotting was problematic for precipitation in an early version of figure 1b. A compromise caricature uses adjacent values and adds an outlier dot for a maximum or a minimum that is not an adjacent value. The compromise caricature indicates the presence of outliers while hiding outlier multiplicity and location details.

In our boxplot graphical representation, the thick rectangle extends from the 1st to the 3rd quartile, the thin rectangle extends to the extrema, and the vertical median line extends outside the thick rectangle (Carr, 1994b). Comparing adjacent medians can typically be based on the judgment of horizontal distance between line endpoints. Using dots for the median (Becker and Cleveland, 1993) is a viable choice since the area symbol is easy to spot even though enclosed in a rectangle. However, the comparison of neighboring medians may not be as accurate since the distance between points is not the correct distance to judge.

The temperature data for figure 1b are straightforward. The value for each grid cell represents the average over the period 1961 to 1990. The variation of values for each ecoregion is due to the different spatial position of its constituent grid cells. The average annual growing degree days is computed directly from daily temperature. If the average daily temperature is over 50 degrees Fahrenheit, the day counts. The degrees accumulated for days that count is the average daily temperature minus 50 degrees. For example, suppose a cell in Florida had a daily average value of 75 degrees, each day for the whole year. The annual growing degree days would then be $365 * (75 - 50) = 9125$. To provide easier to remember (two-digit) numbers, the scale for figure 1b shows growing degree days divided by 100.

The precipitation data for figure 1b has a thick right tail. A logarithmic transformation helps to pull in the tail, and provides better resolution for small precipitation values. The choice here was to use log base 2, since powers of 2 are familiar to many scientists (Cleveland, 1994).

The names in figure 1b are simply integers. The correspondence between the integers and lengthy text names is shown in figure 1a. There is also an elaborate number scheme for ecoregions that extends to ecoregions at Level III and Level IV, but in the current context the elaborate number scheme contributes little. The plotting of names and color link in figure 1b differs from that in Carr et al. (1998). Here the names appear as right aligned numbers and the color link appears after the name. In Carr et al. (1998) color links appear to the left of the left-aligned alphabetic names. Putting the color link near the name speeds the matching process, particularly in going between the name and the ecoregion. However, the name is also linked to the color of the boxplots by position. Showing a colored rectangle or dot next to the name is not necessary and can be omitted by those that find it distracting. We include the extra link here because it makes the color linking more explicit.

As for sequential pattern in the LM plots, the sorting by increasing median temperature arranges the ecoregions in the micromap panel sequence so that northern ecoregions appear in the top panels and the southern ecoregions appear in the bottom panels. This matches common expectation. However there are many details that most people would not know. For example, ecoregion 16 (bottom panel orange) tends to be further north than ecoregion 17 (next to bottom panel in purple) but is generally warmer. The variation for ecoregion 16 is large so not all the region is warmer.

The growing degree days boxplots parallel the temperature boxplots both in terms of mean and variability. However there are some variations in the ranking by median. For example the ranks of ecoregions 1 and 3 in the top panel are switched. Ecoregion 4, the Marine West Coast Forests, at the bottom of the second panel is an anomaly showing many fewer growing degree days than would be expected from the temperature boxplot. This result might differ if the 50-degree base temperature was lowered to 40 or 45 degrees F.

The relation between precipitation median and temperature median does not appear to be strong. Of course juxtaposed medians, bars, or dots provide a poor way to look for a functional relationship even if one variable is sorted. The scatterplot provides the standard for assessing a functional relation between two variables. We proceed to represent the bivariate data for all grid cells and not just the univariate medians.

4.2. LM bivariate boxplots

The scatterplot is a template that is frequently used by scientists. Scientists use it in the context of assessing functional relationships and sometimes for assessing the joint variation of two variables. LM plots include the use of simple scatterplots, with the selected study units shown in the appropriate colors and typically plotted last if the other values are shown.

The current application with 481,475 grid cells warrants designs that handle overplotting. Simple scatterplot solutions include binning the bivariate data into attribute space grid cells and representing the counts. Density estimates and graphical representations also work (Scott, 1992). Carr et al. (1987) were early proponents of binning in context of scatterplot matrices. Their acidic deposition graphic shows spatial selection on a map by point in polygon, color linking, and display of density differences in a scatterplot matrix. This is similar to today's brushing methodology when there is a plot of spatial coordinates as well as attribute values. A notable difference is the computation and display of density differences in the scatterplot matrix. Standard brushing methodology just changes the color of points and (over) plots them in some order. Even if the selected set of points is plotted last, the viewer has the task of comparing densities. Point density assessment has low perceptual accuracy even when one can see the points (Cleveland and McGill, 1984). Density representations are in order.

Simple binning provides density estimates that scale well with data set size. Carr (1998), for example, shows a scatterplot matrix density image representing over 1/2 billion point pairs. The source was a 7K by 8K multispectral satellite image of the Willamette Valley. The bivariate density estimates vary over many orders of magnitude. This makes appreciating the density surface details a time consuming task. Perspective views of bivariate density surfaces in an interactive setting can work when options include change of viewpoint, rescaling and density clipping. For current purposes we seek something far simpler to grasp, a simple bivariate boxplot.

Bivariate boxplots have been little used. One reason is the lack of consensus concerning the definition and calculation of a bivariate median. Carr (1991) proposes hexagon bin boxplots and an easily computed median. The computational sequence bins the bivariate data using a hexagon grid, determines the high-density hexagon cells that contain 50% of the observations, smoothes the high density cells (an optional step), and then erodes the cells using gray level erosion. The gray level erosion process removes counts from the 50% contour cells proportional to the number of exposed hexagon edges. For example, in one erosion cycle a completely surrounded 50% contour cell would have no counts removed, while an edge cell with three exposed edges would have three counts removed. When a cell is emptied it exposes the edges of its neighbors. The process repeats. The last cell eroded gives the location of the median in the binned density representation. This simple procedure generalizes to other regular grids in two and higher dimensions.

Carr (1991) shows a two-way layout of bivariate boxplots. The layout includes the direct display of boxplot differences between adjacent boxplots. For example, an arrow shows the change of the median cell. Those bivariate boxplots were rescaled to focus on the 50% high-density regions and the medians. The direct display of differences facilitates detailed comparison. In our simple example, we omit the difference panels but include the region of bivariate support, just as we show the thin box covering the extrema in 1-D boxplots.

Technically the 50% high-density bivariate region is not exactly analogous to the interquartile range in univariate boxplots. That is, the interquartile range may include regions of low density. The bivariate boxplots here are just one option for a caricature. An outside to inside constrained erosion procedure may provide a better analogue to the univariate boxplot. Some researchers may prefer to emphasize density and show local modes in the high-density region rather than computing some form of median. The important thing is get on with the graphics rather than wait for years seeking consensus on the best of reasonable caricatures.

Figure 2 shows a bivariate boxplot of the combined data in the lower right corner. The purple shows the support of the bivariate pairs, (30-year average annual temperature, 30-year average precipitation on a log scale). The yellow cells are the 50% high-density cells and the black cell is the median cell. The positive association of the high-density cells contrasts strongly to the negative association suggested by the purple region of support. The latter is all one would see in an view with overplotting. How useful it is to study this aggregate plot can be argued. Scientists tend to be interested in detail. When scientists see a plot splitting into non-homogeneous pieces, as in figure 2, they naturally want to study the pieces. Overviews are often valuable for providing a frame of reference for studying the pieces.

The lack of homogeneity and many other complications in the analysis of large data sets have long been identified as an issue. For example, Hall (1980) writes "We can't really expect large data (sets) to be homogeneous." One challenge is "to intelligently cut up the data to increase homogeneity in the parts." Carr (1980) mentions homogeneity issues in a complexity framework involving "vector valued relationships, hierarchical structures, and relational networks" that apply to the same data as the "statistician adopts different viewpoints." (For more recent discussions on massive data sets see, Carr, 1996; CATS, 1996; and Huber, 1994) Given the recurrent use of hierarchical structures in human organizational activities it comes as no surprise that Omernik adopts a hierarchical approach in partitioning North America into the progressively more homogeneous life supporting regions (networks). That his ecoregion hierarchy is not strictly nested reflects a view that phenomena need not necessarily be consistent with cognitively convenient patterns. The incredible complexity and diversity of life on the surface of the earth makes the definition of ecoregions a daunting challenge. That a human mind can structure such a vast amount of domain information and knowledge is amazing.

We mention only a few patterns in figure 2. Ecoregions 15, 18, and 16 (top to bottom order) catch the eye with relatively big bivariate support regions (gray) that have negative slopes. The high density yellow cells also have negative slopes indicating the association of increased rainfall and decreased temperature. An natural query is to see if the association is partly related to altitude. The high density cells in ecoregion 16 show a bifurcation that motivates further investigation and suggests possible subdivision of the ecoregion. However, we will soon see that ecoregion 16 is relatively homogeneous in terms of land cover. Ecoregions 10 and 9 show high positive slope, suggesting that most of the variation is in terms of temperature. Note that both ecoregions cover a large range of latitude. In general, the temperature variation in the bivariate boxplots appears associated with latitude variation in the micromaps. This motivated putting temperature on the y-axis.

Figure 2 uses the univariate median temperature as the basis for sorting, but the micromap layout is different than in figure 1. Due to the size of bivariate boxplots, figure 2 shows four or fewer ecoregions per micromap. The vertical list of names and color links are absent since they are mostly redundant with the names and color tags in each boxplot panel. The list would clarify the bivariate boxplot order that is left to right, top to bottom. Careful comparison of bivariate median temperatures against the grid reveals that the bivariate temperature ranking parallels the univariate median ranking with few exceptions. As indicated earlier, different univariate and multivariate sorting provides the opportunity to bring out other patterns.

4.3. LM time series plots

The boxplots in figure 1b consume relatively little vertical space and height only encodes the class distinction between visual elements representing median, quartiles, and extrema.

With times series plots, the vertical dimension encodes a continuous variable and more resolution is desirable. The simple solution in figure 3 is to overplot the time series for each perceptual group using the same scale. Color time series examples can be found on our web site, in Carr et al. (1998) and in Carr et al. (1997).

Restricting LM plots to gray scale is a bit of a challenge. While the eye is very sensitive to lightness (value or brightness in some articles), it is not easy to quickly discriminate more than a few well-separated levels of lightness. The layout for figure 3 is a 4-4-5-4-4 pattern that necessitates only five distinctions for ecoregions rather than the six in figure 1b. Reducing to only 4 ecoregions for a perceptual group as in figure 2, has merit in terms of easier discrimination, but we chose to show a different layout as our third example. The use of gray level to distinguish among ecoregions motivates us to use a thin black line to show the grids. The time series do overplot. The sorting order that uses the median of monthly medians is helpful for interpretation purposes but not helpful in terms of reducing the overplotting. Substantial overplotting makes it hard to follow the times series, even when using distinct hues. With just a few times series, the reader can usually infer values for hidden lines, but this is extra work. To address the overplotting we used lines of different thickness and indicated this as part of the name and color link. We plotted the thickest lines first so the thinner lines would be visible.

The motivation for figure 3 is that seasonal variation may be more important for characterizing ecoregions than an annual value. The data are the PRISM average monthly precipitation from 1961 to 1990. The value plotted is the median of the grid cells values in each ecoregion. A fancier plot could show staggered vertical boxplots of the monthly values for each of the ecoregions. Ecoregions are sorted by median precipitation.

Several patterns are apparent. The micromaps show that the median of monthly median values tends to be low in the West. The bottom two micromaps show regions in the East except for Ecoregion 4. The time series show some distinctive patterns. Ecoregions 18, 3 and 4 have higher precipitation in winter than during summer. Ecoregion 21 has very high values in June and September. Adjacent ecoregion 4 shows similar precipitation with high values in July and August. Cluster based sorting may help to facilitate presentation of similar seasonal patterns at the possible expense of more overplotting.

4.4. *LM line height plots*

Line height plots can also be considered bar plots. The distinction concerns the width of the bar, with a thin width being used when there are many bars to show in a relatively small area. Figure 4 uses line heights to display 159 values for each ecoregion. The values represented are the area percents of each ecoregion that have been classified as belonging to one of 159 vegetation classes. The percents for each ecoregion add up to one hundred. Many percents are zero or close to zero and are masked by the panel outline. The figure grid lines help in assessing and comparing the larger percents. Representing many variables in LM plots is not much of a problem when there is a common scale.

A big problem with static LM line height plots is labeling. Zooming and clicking can provide a solution in an interactive setting. Figure 4 deals with the problem by limiting labels to broad classes of land cover: agriculture, rangeland, forest, wetland/water, and barren/tundra.

Sorting is of particular interest in this example. The patterns appearing in an unsorted matrix of 21 ecoregions by 159 row percents can appear intimidating and discourage further study. Carr and Olsen (1996) promote a non-traditional multivariate sorting approach that substantially simplifies two-way layout appearance in many applications. The sorting uses a breadth traversal of a minimal spanning tree as defined by Friedman and Rafsky (1979). The motivating notion is that some subtrees of the minimal spanning tree have multivariate points that are relatively close together and that the traversal will place these points together creating small perceptual groups.

In figure 4 the ecoregions are first sorted by traversal of the spanning tree in 159 dimensions (two lines of code in S-PLUS^{TM}). Then we sort columns within each set of land cover variables, such as rangeland, using a spanning tree in the 21 dimensions of different ecoregions. The percents across ecoregions do not add to one but the units are all percents. Carr and Olsen (1996) show a side by side comparison of sorted versus unsorted line height plots for this land cover data using a prior definition of ecoregions. The clustering of bars to similar height bars and white space to white space makes the sorted plot substantially simpler. Simple appearance can be crucial in encouraging scientists to study the patterns. Other sorting can be helpful such as median sorts, first principal component sorts, and clustering dendrogram sorts. We have not conducted a comparative study, but our experience suggests that the minimal spanning tree breadth traversal is particularly effective for producing simple appearance.

As for a brief description, a few forest types dominate ecoregion 8. Ecoregion 9 has several forest types similar to ecoregion 8 but has more agriculture. Ecoegion 21 has several types of wetlands. Ecoregion 14 is very homogeneous in land cover being mostly rangeland with over 40% as one type. In contrast, ecoregion 12 is very heterogeneous. A reasonable guess is that Level 3 ecoregions will substantially subdivide Ecoregion 12 and not Ecoregion 14.

4.5. Glyph plots and a precipitation trend example

Carr and Pierson (1996) favored LM plots over choropleth maps in terms of representing statistical summaries more faithfully. Their reasons include better perceptual accuracy of extraction and easy visual representation of uncertainty. However, classed choropleth maps are not the only way to represent spatially indexed statistical summaries; glyph plots also provide a viable alternative for representing summaries.

Glyph plots provide a common way of presenting statistical summaries on maps. Chambers et al. (1983) note that meteorologists routinely use weather map symbols to simultaneously represent cloud cover, wind direction, and wind speed. Carr et al. (1992) use a less complex glyph for showing simple trends, a ray on top of a gray filled circle sector. In their example, the ray angle represents the sulfate deposition trend and the sector represents 90 percent confidence bounds. A local scale, a circle with lines at regular angles, provides the basis for judging ray angle and the two angles defined by the straight edges of the filled sector. Thus the encoding uses position along a scale, albeit an angular scale. Figure 5 extends the glyph by adding two lines that define two additional angles. The figure 5 glyph shows five angles and is suitable for representing five values in a boxplot summary.

The angular limits of the boxplot glyph in figure 5 are from −135 degrees to 135 degrees. There is some advantage in restricting the limits to 90 degrees so, for example, the local

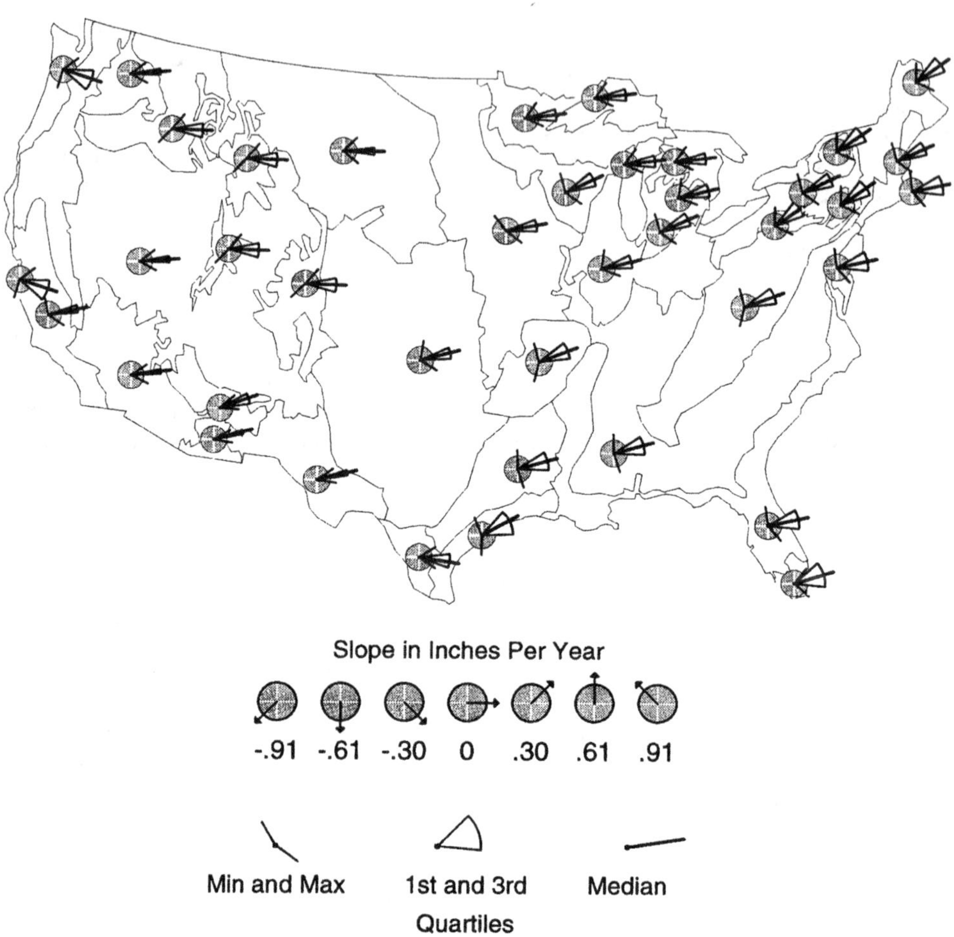

Figure 5. 1961–1996 precipitation trends angular boxplots for stations within ecoregions.

y-coordinate indicates the largest increase. The extension here is to increase the angular resolution.

Other glyph variations are also worth consideration. The variation that we call the Portuguese man-of-war, shortens the extrema lines to the reference circle, cuts out the sector between the extrema, and drops the arc between the quartiles. The gestalt effect is striking, but that does not necessarily help judging the angles accurately. An example appears on the web site.

Since figure 5 borrows from Carr et al. (1992), a few comments about their example seems appropriate before discussing the figure directly. First, their example is notable because the original glyph represents both the estimate and the quality of the estimate using a continuous scale. While continuous representations of means and some form of error bars are routine in statistical graphics, representing uncertainty in mapping contexts is rare. Of course members of the GIS, cartographic, geographic communities are concerned about uncertainty of both spatial coordinates and attribute measures (Goodchild and Gopal, 1989). MacEachren (1994) has made suggestions concerning visual representation of uncertainty on maps. However, routine practice lags behind the concern. There are a few indications of change. For example, The *Atlas of United States Morality* (Pickle et al., 1997) represents estimate quality as a binary variable, good or questionable.

Figure 5 shows boxplots of weather station precipitation trend for the ecoregions over the period 1961 to 1996. We computed the trend as indicated in Section 2. Figure 5 represents weather station trends within each ecoregion using an angular boxplot as indicated above. The boxplot provides a caricature for a batch of data or estimates. There is no assumption that the estimates are independent and identically distributed. The calculation of a single summary trend with confidence bounds for each ecoregion would require sophisticated spatial-temporal modeling. (The confidence bounds in Carr et al. (1992) were the nonparametric bounds on the median for a single station.) The boxplots provide a graphical display of spatial diversity. The boxplot glyph is lacking in that it does not show the number of stations involved for each of the ecoregions. The number of weather stations varies from a low of 12 in ecoregion 21 to 614 in ecoregion 7. While the area of the ecoregions and other factors must be taken into account, it seems safe to say that monitoring of ecoregions was not a major factor in weather station placement.

Figure 5 shows some patterns. Horizontal median lines suggest no change. Most of the median lines suggest an increase with larger increases in the Northeast. The Pacific Northwest rain forest appears to show decreasing precipitation. The white sectors with black outlines show the interquartile range. Some of these are tight about median and most are fairly symmetric about the median. The extrema are often interesting if only as a check. The legend suggests the possibility of changes from −.91 to .91 inches per year. Several glyphs appear with extrema near the minimum.

The existence of multiple polygons for the same ecoregion complicates examination of figure 5. The same glyph appears for each polygon that comprises a given ecoregion. This gives more visual weight to ecoregions with multiple polygons. If unequal visual weight is given, a more reasonable choice is to favor large area ecoregions. The glyph placement in figure 5 avoids glyph overplotting. Symbol congestion control becomes increasingly problematic as the number of polygons increase, for example with Level III ecoregions. LM plots avoid multiple glyph, placement, and symbol congestion problems and the linear scale boxplot is a bit easier to assess than an angular scale boxplot.

Some glyphs are designed to have good perceptual accuracy of extraction (see the framed-rectangles in Cleveland and McGill, 1984). Like perspective views of a surface, some glyph representations provide a spatial gestalt impression that is hard to match with other representations. Thus it can be of benefit to study both LM plots and glyph plots.

5. LM plot history, connections to other research and challenges

We claim that the LM template is new, but there are, of course, many connections to previous graphics and conceptualizations. While we were intrigued by the thumbnail images of Eddy and Mockus (1996), a stronger connection is to the work of Edward Tufte. The LM plots belong to a class of graphics that Tufte (1983, 1993, 1997) calls small multiples. In *The*

Visual Display of Quantitative Information, his eloquent description of well-designed small multiples include phrases such as "inevitably comparative", "deftly multivariate", "efficient in interpretation", and "often narrative in content." We designed LM plots with the hope that such phrases would apply. In *Visual Explanations*, Tufte calls particular attention to the explanatory power of parallelism. While our use of parallelism precedes this book, Tufte's earlier examples may well have guided us to make parallelism a fundamental part of the LM plot design.

LM plots emerged as a way of linking row-labeled plots (Carr, 1994a, b, c; Carr and Nusser, 1995; Carr and Olsen, 1996; and Carr, et al., 1996) to maps. The row-labeled plots in turn build upon work of Cleveland (1985) and Cleveland and McGill (1984). In fact, the development of row-labeled plots was part of an effort to encourage U.S. EPA staff to use Cleveland's dotplots in EPA graphics. Dissatisfaction with the look of early S-PLUS dotplots (Cleveland, 1993b) and the promise of multiple panel layouts for expressing complex tables as plots led to the development of new S-plus functions (Carr, 1994a and Carr, 1997). While the row-labeled plot development was independent of the Trellis™ graphics development, there are similarities. This is not surprising, since Cleveland's design ideas were important in both and S-PLUS was a common computing environment. While LM plots may not incorporate all of Cleveland's guidance, (for example space and map size constraints have so far taken precedence over banking), we strive to put his guidance to work in our designs.

The linking of maps and statistical graphics also builds upon the work of Monmonier (1988) who connected contemporary methodology from cartographic and statistical graphics communities. One of his many interesting examples has a map on the left, labels in the middle and bar plots on the right. This example is a precursor to our LM plots. We have followed Monmonier's work over the years. The state map caricature in Carr and Pierson (1996) was specifically inspired by Monmonier (1993) and adapted from coordinates that he graciously supplied.

Plot production convenience remains a big issue. If LM plots are to be used they need to be easily produced. The general S-PLUS tools we developed (anonymous ftp to galaxy.gmu.edu and change directory to pub/dcarr/newsletter/lmplots) are flexible building blocks but not easy push button tools. The software includes a Visual Basic front end developed to simplify production of LM dot plots and time series plots for states. This is just a start toward simple production. Much work remains to design micromaps for new applications and to develop software that makes it easy to produce a wide range of LM plots. Much research is appropriate concerning compromises and variations that are motivated by plot purpose, audience, specific data and metadata.

6. Acknowledgments

S-PLUS is a registered trademark. Trellis is a trademark of Lucent Technologies. We thank D. Andrew Carr for creating Visual Basic front ends for state LM dotplots and time series plots and for addressing layouts with different numbers of study units. EPA funded the majority of the work behind this paper under cooperative agreements No. CR8280820-01-0 and No. CR825564-01-0. Additional federal agencies, BLS and NCHS, supported some facets of this work. The article has not been subject to review by BLS, and NCHS so does not necessarily reflect the view of the agencies, and no official endorsement should be inferred. It has been subjected to EPA's peer and administrative review and approved for publication. The conclusions and opinions are solely those of the authors and are not necessarily the views of the Agencies.

References

Artic Monitoring and Assessment Programme 1997. Arctic Pollution Issues: A State of the Arctic Environmental Report. AMAP, Box 8100 Dep. N-0032 Oslo, Norway.

Bailey, R.G. 1995a. Description of the ecoregions of the United States. USDA Forest Service, Washington, DC, Misc. Publ. No. 1391 (rev.).

Bailey, R.G. 1995b. Ecosystem Geography. New York: Springer-Verlag.

Bailey, R.G. 1998. Ecoregions map of North America: Explanatory Note. USDA Forest Service, Washington, DC, Misc. Publ. 1548.

Becker , R.A. and Cleveland, W.S. 1993. Discussion of graphical comparison of several linked aspects: Alternative and suggested principles. J. Computational and Graphical Statistics 2:41–48.

Bertin, J.B. 1983. Semiology of Graphics Diagrams Networks Maps. Berg, W.J. (translator). London England: The University of Wisconsin Press.

Brewer, C.A. 1997. Spectral schemes: Controversial color use on maps. Cartography and Geographic Information Systems 24:203–220.

Carr, D.B. 1980. Impact of large on the analysis process. D.L. Hall (Ed.), ALDS 1979 Panel Review, Pacific Northwest Laboratory, Richland, WA, PNL-SA-8781.

Carr, D.B. 1991. Looking at large data sets using binned data plots., A. Buja and P. Tukey (Eds.), Computing and Graphics in Statistics, New York: Springer-Verlag, 7-39.

Carr, D.B. 1994a. Converting plots to tables. Center for Computational Statistics, George Mason University, Fairfax, VA, Technical Report No. 101.

Carr, D.B. 1994b. A colorful variation on boxplots. Statistical Computing & Graphics Newsletter 5(3):19–23.

Carr, D.B. 1994c. Using gray in plots. Statistical Computing & Graphics Newsletter 5(2):11–14.

Carr, D.B. 1996. Perspectives on the analysis of massive data sets. Proceeding of the 27th Symposium on the Interface, Interface Foundation of North America, Fairfax, VA, Computing Science and Statistics 27:410–419.

Carr, D.B. 1997. Some simple Splus tools for matrix layouts. U.S. Bureau of Labor Statistics, Washington, DC, Statistical Note Series No. 42.

Carr, D.B. 1998. Multivariate Graphics. P. Armitage and T. Colton (Ed.), Encyclopedia of Biostatistics, New York: John Wiley and Sons, 4:2864–2886.

Carr, D.B., Littlefield, R.J., Nicholson, W.L., and Littlefield, J.S. 1987. Scatterplot matrix techniques for large N.J. American Statistical Association 82:424–436.

Carr, D.B. and Nusser, S.A. 1995. Converting tables to plots, a challenge from Iowa State. Statistical Computing & Graphics Newsletter 6(3):11–18.

Carr, D.B. and Olsen, A.R. 1996. Simplifying visual appearance by sorting: An example using 159 AVHRR classes. Statistical Computing & Graphics Newsletter 7(1):10–16.

Carr, D.B., Olsen, A.R., Courbois, J.P., Pierson, S.M., and Carr, D.A. 1998. Linked micromap plots: Named and described. Statistical Computing & Graphics Newsletter 9(1):24–32.

Carr, D.B., Olsen, A.R., and White, D. 1992. Hexagon mosaic maps for display of univariate and bivariate geographical data. Cartography and Geographic Information Systems 19: 228–236, 271.

Carr, D.B. and Pierson, S. 1996. Emphasizing statistical summaries and showing spatial context with micromaps. Statistical Computing & Graphics Newsletter 7(3):16–23.

Carr, D.B., Somogyi, R., and Michaels, G. 1997. Templates for looking at gene expression clustering. Statistical Computing & Graphics Newsletter 8(1):20–29.

Carr, D.B., Valliant, R., and Rope, D. 1996. Plot interpretation and information webs: A time-series example from the Bureau of Labor Statistics. Statistical Computing & Graphics Newsletter 7(2):19–26.

CATS. 1996. Massive data sets—Proceedings of a workshop. Committee on Applied and Theoretical Statistics, National Research Council, National Academy Press, Washington, DC.

Chambers, J.M., Cleveland, W.S., Kleiner, B., and Tukey, P.A. 1983. Graphical Methods for Data Analysis. Pacific Grove, CA: Wadsworth and Brooks/Cole.

Cleveland, W.S. 1993a. Visualizing Data. Summit NJ: Hobart Press.

Cleveland, W.S. 1993b. A model for studying display methods of statistical graphics. J. Computational and Graphical Statistics 2(4):322–343.

Cleveland, W.S. 1994. The Elements of Graphing Data. Summit, NJ: Hobart Press.

Cleveland, W.S. and McGill, R. 1984. Graphical perception: Theory, experimentation, and application to the development of graphical methods. J. American Statistical Association 79:531–554.

Daly, C., Neilson, R.P., and Phillips, D.L. 1994. A statistical-topographic model for mapping climatological precipitation over mountainous terrain. J. Applied Meteorology 33:140–158.

Dent, B.D. 1990. Cartography—Thematic Map Design. Dubuque, IA: Wm C. Brown Publishers.

Dorling, D. 1995. A New Social Atlas of Britain. Chichester, England: John Wiley & Sons.

Eddy, W.F. and Mockus, A. 1996. An interactive icon index: Images of the outer planets. J. Computational and Graphical Statistics 5(1):101–111.

Eick, S.G. 1996. Visualizing large data sets. Massive Data Sets —Proceedings of a Workshop, Committee on Applied and Theoretical Statistics, National Research Council, National Academy Press, Washington, DC.

Friedman, J.H. and Rafsky, L.C. 1979. Multivariate generalizations of the Wald-Wolfowitz and Smirnov two-sample tests. Annals of Statistics 7(4):697–717.

Frigge, M., Hoaglin, D.C., and Iglewicz, B. 1989. Some implementations of the box plot. The American Statistician 43:50–54.

Goodchild, M. and Gopal, S. (Ed.). 1989. Accuracy of Spatial Databases. New York: Taylor & Francis.

Grant, J.P. 1993. The State of the World's Children 1993. Oxford: Oxford University Press.

Green, S., Marchionini, G., Plaisant, C., and Schneiderman, B. 1997a. Previews and overview in digital libraries: Designing surrogates to support visual information-seeking. Human Computer Interaction Laboratory, University of Maryland, College Park, MD, Technical Report 97-16.

Green, S., Tanin, E., Plaisant, C., Schneiderman, B., and Olsen, L. 1997b. The End of zero-hit queries; Query preview for NASA's Global Change Master Directory. Human Computer Interaction Laboratory, University of Maryland, College Park, MD, Technical Report 97-20.

Hall, D.L. 1980. Large data sets analysis concepts. D.L. Hall (Ed.), ALDS 1979 Panel, Pacific Northwest Laboratory, Richland, WA, PNL-SA-8781.

Huber, P.J. 1994. Huge data sets. R. Dutter and W. Grossman (Ed.). Compstat 1994: Proceedings, Heidelberg: Physica Verlag.

Julesz, B. 1986. Texton gradients: The texton theory revisited. Biological Cybernetics 54:245–251.

Kosslyn, S.M. 1994. Elements of Graph Design. New York: W. H. Freeman and Co.

Loveland, T.R., Merchant, J.W., Reed, B.C., Brown, J.F., Ohlen, D.O., Olson, P., and Hutchinson, J. 1995. Seasonal land cover regions of the United States. Annals Association of American Geographers 85:339–355.

MacEachren, A.M. 1994. Some Truth with Maps: A Primer on Symbolization & Design. Washington, D.C.: Association of American Geographers.

MacEachren, A.M. 1995. How Maps Work—Representation, Visualization, and Design. New York: The Guilford Press.

McGill, R., Tukey, J.W., and Larsen, W.A. 1978. Variation of boxplots. The American Statistician 32:12–16.

Monmonier, M. 1988. Geographical representations in statistical graphics: A conceptual framework. 1988 Proceedings of the Section on Statistical Graphics, American Statistical Association, Alexandria VA. pp. 1–10.

Monmonier, M. 1991. How to Lie with Maps. Chicago: The University of Chicago Press.

Monmonier, M. 1993. Mapping It Out. Chicago: The University of Chicago Press.

Olsen, A.R., Carr, D.B., Courbois, J.P., and Pierson, S.M. 1996. Presentation of data in linked attribute and geographic space. Poster presentation, 1996 Joint Statistical Meetings, Chicago, IL.

Omernik, J.M. 1987. Ecoregions of the conterminous United States. Annals Association of American Geographers 77(1):118–25.

Omernik, J.M. 1995. Ecoregions: a spatial framework for environmental management. W.S. Davis, and T.P. Simon (Ed.), Biological Assessment and Criteria: Tools for Water Resource Planning and Decision Making. Boca Raton, FL: Lewis Publishers. pp. 49–62.

Pickle, L.W., Mingle, M., Jones, G. K., and White, A.A. 1997. Atlas of United States Mortality. Hyattsville, MD: National Center for Health Statistics.

Plaisant, C., Mushing, R., Snyder, A., Li, J., Heller, D., and Schneiderman, B. 1998. LifeLines: Using visualization to enhance navigation and analysis of patient records. Human Computer Interaction Laboratory, University of Maryland, College Park, MD, Technical Report 98-80.

Scott, D.W. 1992. Multivariate Density Estimation, Theory, Practice and Visualization. New York: John Wiley & Sons.

Tufte, E.R. 1983. The Visual Display of Quantitative Information. Cheshire, CT: Graphics Press.

Tufte, E.R. 1990. Envisioning Information. Cheshire, CT: Graphics Press.

Tufte, E.R. 1997. Visual Explanations. Cheshire, CT: Graphics Press.

Tukey, J.W. 1993. Graphic comparisons of several linked aspects: Alternatives and suggested principles. J. Computational and Graphical Statistics 2:1–33.

USGS. 1985. 1 : 250,000-scale Digital Elevation Models (DEM) also known as 1-Degree DEM's. U.S. Geological Survey, Washington, DC.

Wahlström, E., Hallanaro, E., and S. Manninen. S. (Ed.). 1996. The Future of the Finnish Environment. Helsinki: Edita Ltd.

Wiken, E.B. 1986. Terrestrial ecozones of Canada. Environment Canada, Ottawa, Ontario, Canada, Ecological Land Classification Series No. 19.

Woods, D. 1992. The Power of Maps. New York: The Guilford Press.

Zawitz, M.W. (Ed.). 1983. Report to the Nation on Crime and Justice, The Data. U.S. Department of Justice, Washington, DC.

Daniel B. Carr is a professor in the department of Applied and Engineer Statistics at George Mason University. He received his Ph.D. in Statistics from the University of Wisconsin, Madison in 1976. He is Fellow of the American Statistical Association and the Washington Academy of Science. His research interests include information and knowledge visualization, exploratory data analysis and datamining for large data sets, data analysis management and environments for depth of thought, the human computer interface and web distribution of statistical summaries.

Dr. Anthony R. Olsen is a senior mathematical statistician at the U.S. Environmental Protection Agency's Western Ecology Division, Corvallis, Oregon. He received a Ph.D. in statistics from Oregon State University in 1973. He has received the Distinguished Achievement Award from the American Statistical Association's Section on Statistics and the Environment and the distinguished statistical ecologist award of the International Association for Ecology. Dr. Olsen's research focuses on the development of large-scale ecological monitoring programs and statistical graphics for geographical data.

Suzanne M. Pierson is a GIS analyst/cartographer for OAO Corporation, contractor to the US EPA in Corvallis, OR. She graduated from Oregon State University in 1985. Her research interests include developing simplified graphics and maps from complex scientific data.

Jean-Yves Pip Courbois is a student at Oregon State University nearing completion of a Ph.D. in Environmental Statistics. His research includes statistical graphics and linear models for finite populations.

Data Mining and Knowledge Discovery, 4, 69–80 (2000)

Visual Data Mining in Atmospheric Science Data

MÁRCIA MACÊDO macedo@iastate.edu
Department of Statistics, Iowa State University, 102 Snedecor Hall, Ames, IA 50011-1210, USA

DIANNE COOK dicook@iastate.edu
Department of Statistics, Iowa State University, 325 Snedecor Hall, Ames, IA 50011-1210, USA

TIMOTHY J. BROWN tbrown@dri.edu
Desert Research Institute, 2215 Raggio Parkway, Reno, NV 89512-1095, USA

Editors: Timothy Brown and Paul Mielke, Jr.

Abstract. This paper discusses the use of simple visual tools to explore multivariate spatially-referenced data. It describes interactive approaches such as linked brushing, and dynamic methods such as the grand tour, applied to studying the Comprehensive Ocean-Atmosphere Data Set (COADS). This visual approach provides an alternative way to gain understanding of high-dimensional data. It also provides cross-validation and visual adjuncts to the more computationally intensive data mining techniques.

Keywords: multivariate analysis, statistical graphics, exploratory data analysis, high-dimensional data, interactive graphics, linked brushing, grand tour

1. Introduction

There is a tendency in data mining, especially on a new warehouse, to throw the most heavy-duty tool available at the data, and expect it to pull out all the interesting information. Interactive data visualization can often be used to identify features in the data that are not revealed by black-box methods. Simple graphics can often be illuminating, especially when enhanced by interaction, and interesting local anomalies—small deviations from the overall patterns—are often easier to spot. This paper describes the use of simple graphical methods for data mining.

There have been significant advances in the scope of graphical tools available for data mining. In particular, there have been major advances in the way a user can interact with plots, and also to make static plots dynamic. The currently available tools open up the world of high-dimensional data to visual inspection. We discuss these methods as they apply to atmospheric science data. The paper does not attempt to expose previously unknown features in our example data but rather to expose the reader to the new methodology and software. These new methods can help data mining in several ways: (1) it is possible to uncover previously unknown features, (2) it is easier and speedier to reach understanding of features, and (3) it is possible to cross-validate conclusions and statements about the data made from other methodologies.

We introduce the graphical approaches by first describing the interactive methods in Section 2, and follow with approaches to dynamic plots in Section 3. Throughout the paper we use the Comprehensive Ocean-Atmosphere Data Set (COADS)[1]. This data is a compilation of in situ weather observations taken by merchant marines over the past 150 years (Elms et al., 1993). Various cleaning and processing was done to get regularly gridded data (Woodruff et al., 1993). In this paper, we examine monthly mean values of sea surface temperature (SST), sea level pressure (SLP), wind speed (WndSpd) and wind direction (WndDir) for the period of January 1980–December 1991, from 30°S to 30°N and 160°E to 75°W. We combined these monthly values into one long-term mean gridded data set for each variable. (Data was re-formatted and organized by Macêdo, 1998.) The figures were produced using the XGobi[2] (Swayne et al., 1998) software and ArcView 3.0[3]. XGobi is software for visualizing high-dimensional data through the manipulation of scatterplots, using interaction such as brushing and identification, and dynamic methods such as the grand tour. There is a seamless interprocess communication link between ArcView 3.0 and XGobi, which allows plot characteristics to be visualized simultaneously within the two packages[4] (Symanzik et al., 1999; Symanzik et al., 1997) in real time.

2. Interactive methods

The key to interaction with plots is to link information in multiple views. The multiple views are usually several different plots of the data, for example, several histograms, several pairwise scatterplots, and a map. There is usually a logical one-to-one relationship between points in one view and those in another. So the most common approach to linking plots is to use one-to-one brushing (coloring or changing the symbol), or identifying points in one view, and observing their location in another view. Color brushing is demonstrated in figures 3–5. (The colors have been translated to gray scale for this paper.)

Linking information between plots has been an active area of research for the last twenty years. Newton (1978) coined the term "brushing" to be interactively painting a group of points using a unique color or symbol (glyph), usually with a rectangular-, circular- or polygonally-shaped "brush". McDonald (1982) introduced the term "linked brushing" for cross-referencing information between plots. Linking information between maps and other types of plots is especially useful for atmospheric science data. Examples can be found in Bao and Anselin (1997), Carr et al. (1987), Dykes (1996), Haining et al. (1996), MacDougall (1992), McDonald and Willis (1987), Monmonier (1989) and Unwin et al. (1990).

Linked brushing effectively allows the user to extract information about the conditional distributions of the multivariate data. For example, with linked brushing it is possible to make rapid queries such as "what is the distribution of sea level pressure given that the sea surface temperature is between 28°C and 30°C?" A good reference for the use of conditional distributions in atmospheric science data is Wilks (1995).

Atmospheric scientists typically examine wind direction in the form of geographic maps overlaid with arrows representing the wind direction. Figure 1 shows an example map of COADS climatological wind vectors used in the linked brushing examples below. Though general features are revealed in this type of plot (e.g., the southeast and northeast trade winds, a large, persistent area of easterly winds, and the Intertropical Convergence Zone

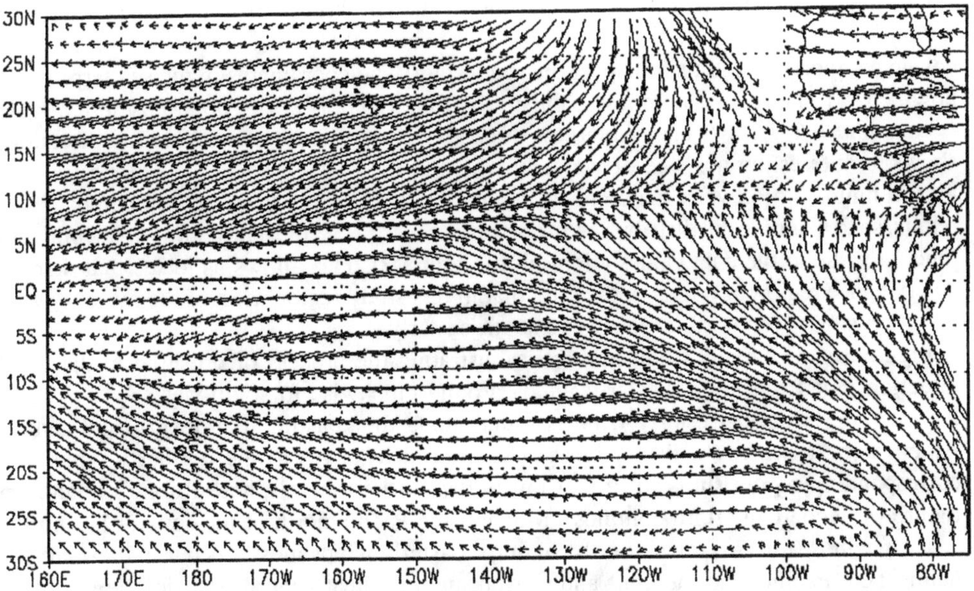

Figure 1. Long-term climatology (1980–1991) of COADS winds used in the linked brushing examples. Vector arrow points towards direction wind is blowing to, and length represents speed (longest vector represents 9 m/s).

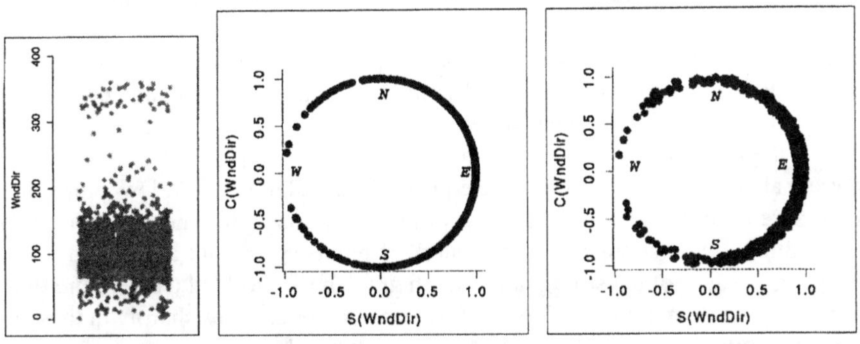

Figure 2. Textured dot plot (left plot) of wind directions (0° to 360°) as used in figure 1. Sine versus cosine plot (middle plot) of wind direction; geographic directions are given as N (north), E (east), S (south) and W (west). Sine versus cosine of wind direction except with "jitter" added (right plot).

(ITCZ) between 5°N and 10°N), it is difficult to visually compare details in one area to another, especially given the large spatial distances. Linked brushing can help digest this information, by allowing us to focus on smaller subsets of wind direction. We show in the next few figures how linked brushing allows dynamic inspection of the relationship between wind direction and geography.

First, it is necessary to organize the wind direction variable into a form suitable for both brushing and interpretability (figure 2). Wind direction is an angle, that is, it is a modular

360 degree measurement (0° = 360°). Looking at this variable as a histogram, or as a textured dot plot (left plot in figure 2) is not ideal, so we convert the measurement into the sine and cosine components and then inspect the scatterplot. The coordinates in the scatterplot can now be read like regular compass points: *top of the circle is north, bottom is south, right is east, and left is west.* The scatterplot of cos(wind direction) versus sin(wind direction) shows an interesting absence of westerly winds in the $(-1, 0)$ part of the circle (middle plot in figure 2). Between the compass points north and south there are an abundance of points, so to examine the distribution in these areas we "jitter" the data adding some uniform random value to the data values (right plot in figure 2). This allows us to see that there is a very dense region of points, indicating that in this study region most of the average wind direction is between north-easterly and south-easterly.

Figure 3 shows many small multiples illustrating the process of brushing in the wind direction plot from north-westerly winds around clockwise to south-westerly winds. In the map views north-westerly wind is found along the coastline of Mexico, and central America (first two rows, first column). As the brush moves to the northerly winds the corresponding region shifts to further out in the northern hemisphere Pacific Ocean (top two rows, second and third columns). Much of the Pacific north of the equator has north-easterly winds (top two rows, fourth column). Easterly winds mostly occur in the southern hemisphere Pacific, and the Gulf and Caribbean areas near central America (bottom two rows, first column). Moving to the south-easterly winds we see that they mostly occur in the southern Pacific (bottom two rows, second column). The southerly winds occur close to the south American coast (bottom two rows, third column), and right on the northern south American coast, around Colombia, the winds are south-westerly (bottom two rows, fourth column).

Figure 4 demonstrates relationships between sea surface temperature and sea level pressure. First, there is a strong negative correlation between temperature and pressure: generally, higher temperature is associated with lower pressure. At higher temperatures (points brushed in the left plot) the concentration is more dense, and points are less spread. These regions are also concentrated around the equator (bottom left plot). At low temperature and high pressure (middle plot) there is more variability amongst the measurements. There is also something a little strange: several "lines" of points spread out at this end of the plot. This appears because of the strong north-south gradients of gridded temperature that occur between 30°S–25°S and 25°N–30°N, that is, temperature changes sharply in these regions in the north-south directions. The right top and bottom plots explore the variability in the cold, moderate pressure region, corresponding to a thin strip of ocean along both Mexican and south American coasts.

Figure 5 illustrates some curious features in the relationship between sea level pressure and wind speed. There are three nodes extending at slow wind speeds, two at low sea level pressures, and one at high sea level pressure. Linked to the map view we can see that the lowest node (left plots, top and bottom) corresponds to a cohesive spatial region in the western Pacific near New Guinea and Australia. The node just above this (middle plots, top and bottom) corresponds to a cohesive spatial region along the south and central American coastlines. The node at high pressure (right plots, top and bottom) corresponds to a cohesive spatial region in the south-eastern Pacific.

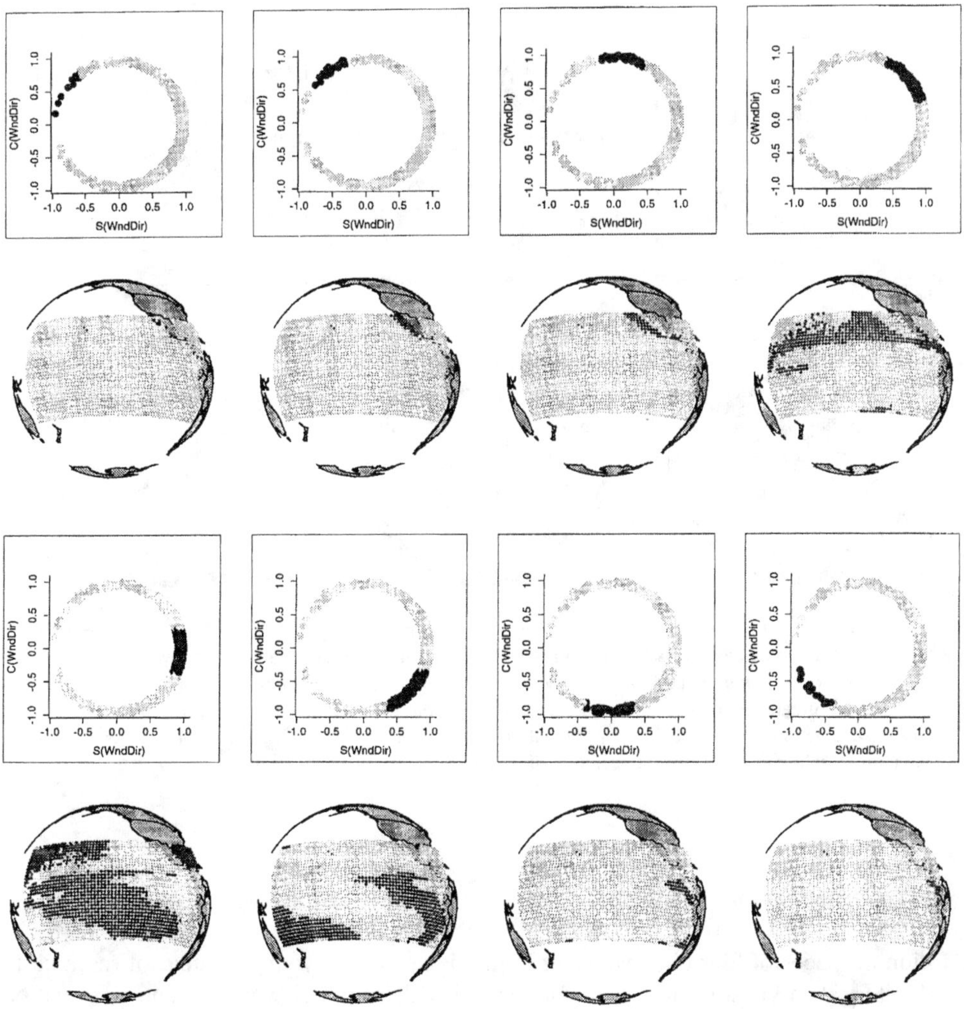

Figure 3. Brushing in the wind direction plot in XGobi, linked to map displayed in ArcView. The brush moves from north-westerly winds clockwise around the circle to northerly winds, then to easterly winds, then to southerly winds, and back to south-westerly winds (top row of plots, and second from bottom row of plots). The second and bottom rows show the corresponding map views.

We have been exploring global or large-scale trends in the data. It is also interesting to explore in finer spatial detail, identifying and examining anomalies where small regions differ dramatically from close neighboring regions, or similarity patterns of neighboring regions. This is best achieved with spatial dependence plots such as the variogram cloud plot. Links between a variogram cloud plot and a map require more complex wiring: a point in the variogram cloud links to two points (represented by a line) in the map. Examples can be found in Cook et al. (1997), Haslett et al. (1991), Unwin et al. (1990).

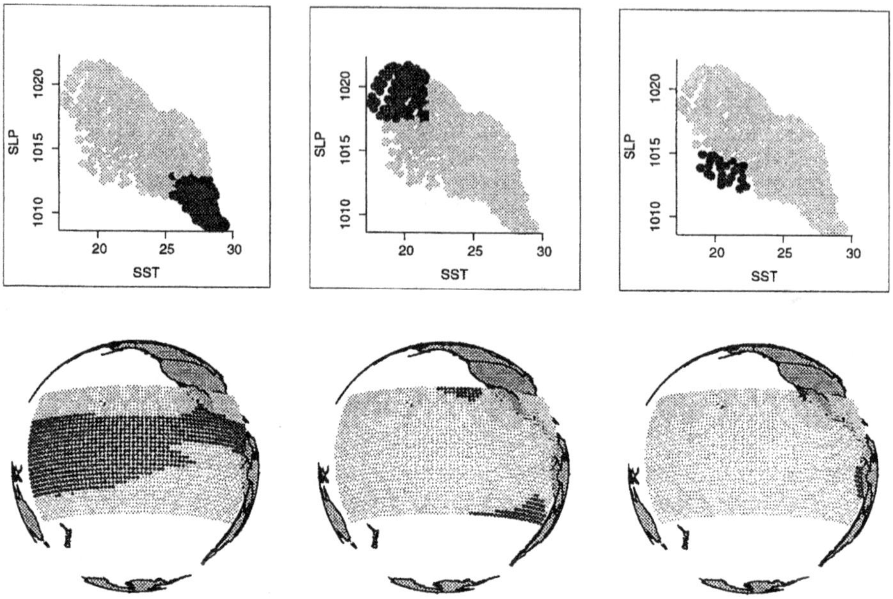

Figure 4. Brushing in the sea level pressure vs sea surface temperature plot in XGobi, linked to the ArcView map. Left top and bottom plots show the variability is least at warm water temperatures and low sea level pressure which corresponds to a wide band around the equator. At cool water temperatures and high sea level pressure the variability is greater, and these correspond to regions in both the north and south portions of eastern Pacific (middle top and bottom plots). The right top and bottom plots explore the variability in the cold, moderate pressure region, corresponding to a thin strip of ocean on Mexican and south American coasts.

3. Dynamic methods

Motion in spaces of four dimensions or higher is one of the youngest areas of research in exploratory data visualization. Dynamic graphics allow the viewer to obtain information about the joint distribution of the data, or the "shape" of the multivariate data. The primary tool is the "grand tour" (Asimov, 1985; Buja and Asimov, 1986), which can be thought of as an extension of 3-D rotation to higher dimensional rotation. Details of the algorithm are available in Buja et al. (1997).

A grand tour is effectively a continuous sequence of low-dimensional projections of the high-dimensional data, much like a computer generated 3-D rotation is a continuous sequence of 2-D projections of 3-D data. In XGobi the tour is a sequence of 2-D projections. It is possible to define tours of 1-D, 2-D or 3-D projections, and this is done in some other software systems. The grand tour allows us to visualize relationships that are more than pairwise in nature, such as clustering, multivariate dependencies and outliers. These features are often missed if one only views histograms and pairwise scatterplots.

There are two major modifications of the grand tour that allow finer control over the sequence of projections. There are guided searches (Cook et al., 1995) and manual control

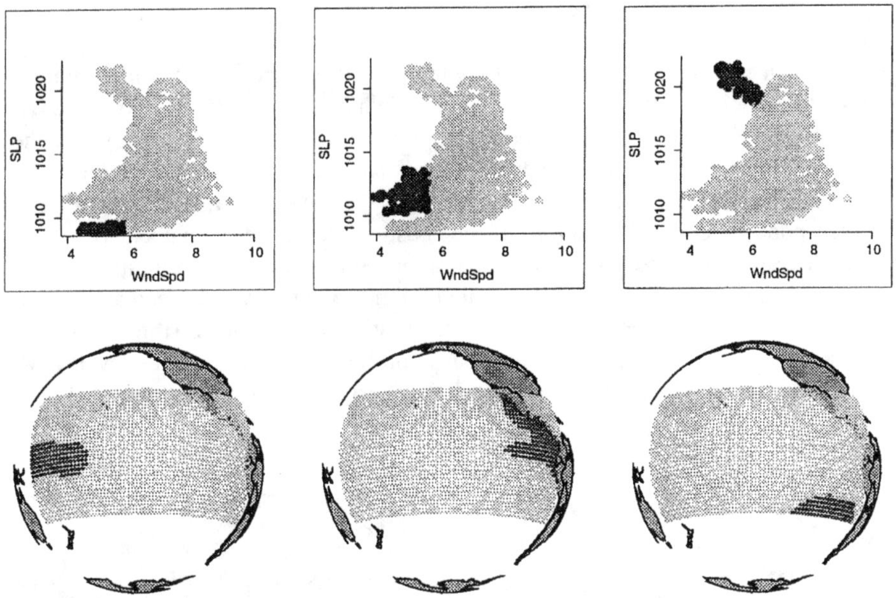

Figure 5. Brushing in the sea level pressure vs wind speed plot in XGobi, linked to the ArcView map. There are some unusual shapes in the plot of sea level pressure and wind speed: three nodes at low values of wind speed, two at low sea level pressures, and one at high sea level pressure.

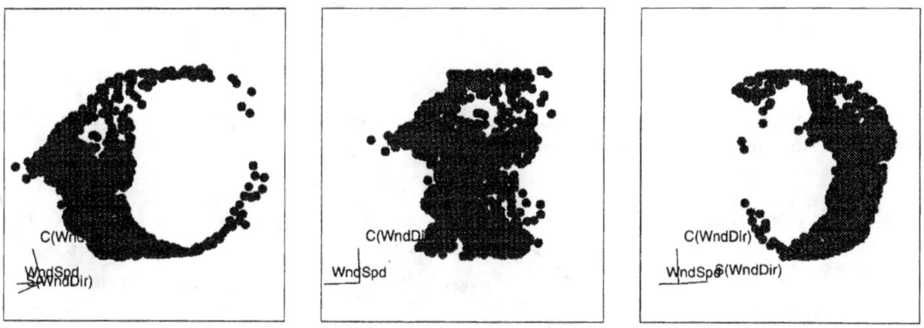

Figure 6. Three 2-D projections of wind speed, cos(wind direction) and sin(wind direction).

(Cook and Buja, 1997). Guided searches provide the viewer with more of the "interesting" views, and manual control allows the user to build in prior knowledge, or refine the view by sharpening the structure or simplifying the interpretation.

The COADS data is best understood with simpler 3-D rotations, especially when inspecting the relationship between wind direction and the other variables. Figure 6 shows several 2-D projections of wind speed plotted against the cosine and sine of wind direction. These points lie on a cylinder in 3-space: the distribution of points in a cylinder is of interest because this allows us to examine the joint distribution of wind speed and wind direction.

Figure 7 shows high wind speed points brushed, which correspond to a cohesive spatial region in the north-central Pacific around Hawaii. Figure 8 shows low wind speed points brushed, which correspond to a region along the central American coast and a small pocket in the western Pacific near New Guinea. The low wind speed values have a range of wind direction values as illustrated if we rotate the plot a little (right plot). If we rotate the plot further, two other unusual features appear. One is an area where the points are sparse and almost gridded in appearance. These points are brushed in figure 9, and it can be seen that they correspond to a cohesive region off the coast of Mexico. The sparseness of points in the variable space compared to the close spatial proximity indicate that in this region of the ocean there is large variability or rapid change in the values of wind speed and direction.

The COADS data can probably be well understood with one, two and three variable plots, so the full power of a high-dimensional tour is probably not needed. However, if we use the grand tour to inspect all four variables—sea surface temperature, sea level pressure,

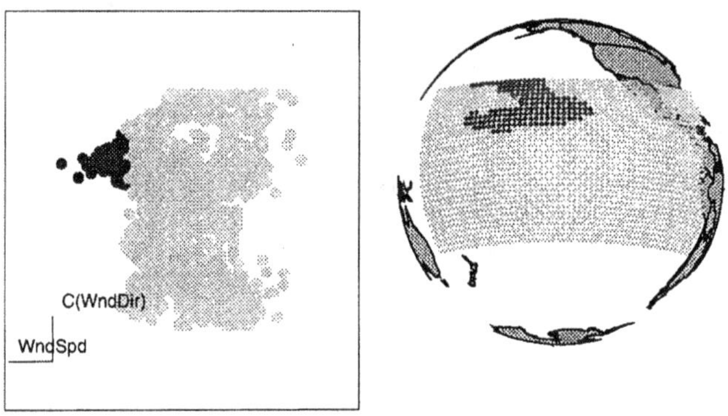

Figure 7. High wind speed values are brushed in the XGobi plot (left) and these correspond to a cohesive spatial region in the north-central Pacific around Hawaii (right plot).

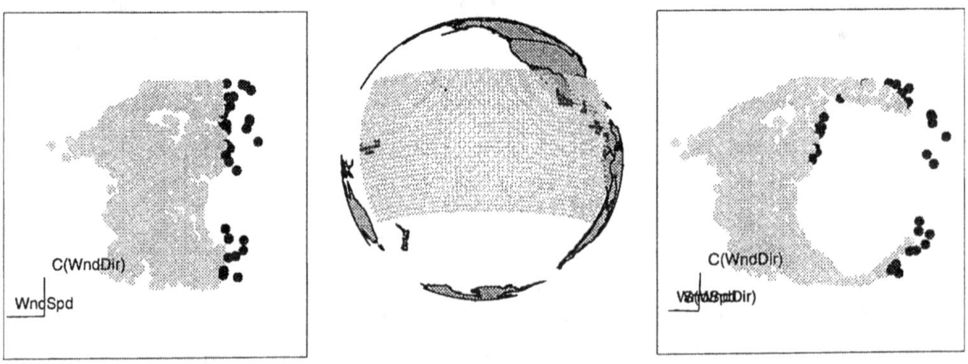

Figure 8. Low wind speed values are brushed in the XGobi plot (left) and these correspond mostly to localized neighborhoods along the central American coast (middle plot). If you rotate the XGobi plot a little, it can be seen that the low wind speeds have considerably different wind directions (right plot).

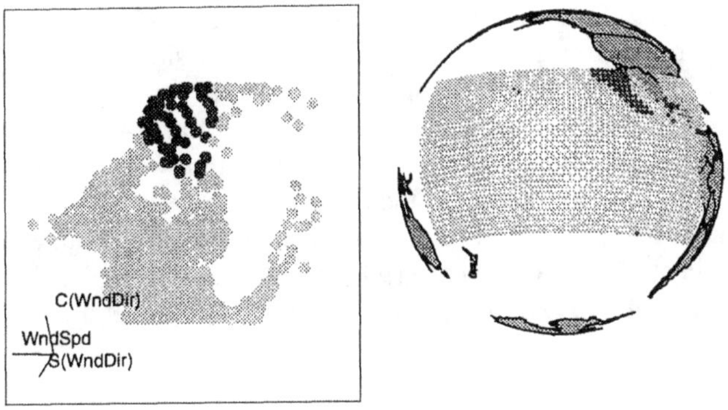

Figure 9. Another rotated view of wind speed, cos(wind direction) and sin (wind direction) revealing a sparse patch of points (brushed). These points correspond to a cohesive spatial region off the Mexican coast (right plot). This indicates that this region has relatively high variability in wind speed and wind direction.

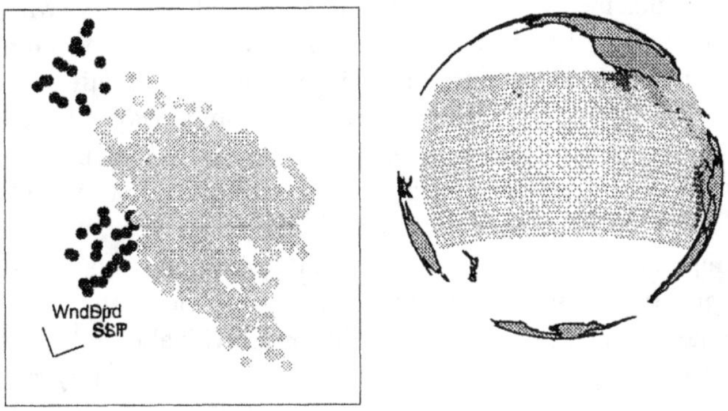

Figure 10. A grand tour view consisting of a projection of four variables—wind speed, wind direction, sea surface temperature and sea level pressure. Two small patches of sparse points can be seen illustrating regions of considerable variability. These correspond to locations off the coast of Mexico and South America.

wind speed and wind direction—a few small features can be seen. Figure 10 illustrates a grand tour view revealing two small patches of larger variation (brushed). These points correspond to two small neighborhoods along the coast of Mexico and South America. The view in the grand tour is a projection that is roughly the average of wind speed and wind direction against roughly the average of sea surface temperature and sea level pressure. The axis in the lower left corner of the XGobi plot indicates this: the length of each axis indicates the magnitude, and the direction of the axis indicates the direction in which the variable contributes to the projection. Interpretation of these axes is similar to interpretation of principal component (PCA or EOF) or factor loadings.

This type of visual inspection also can be used in conjunction with cluster analysis results to extract information about similarities of conditions across the region of interest. Cluster analysis is an exploratory technique used to help "group" observations into like-valued clusters. (See Wilks (1995) for more explanation.) It makes most intuitive sense when the data has neatly separated groups of observations. But in most situations we have data like the COADS example: one large "lump" of points with irregular "protrusions". Cluster analysis can still be helpful here, rather as a way to "partition" or "carve-up" the observations into regions. Cross-checking the output from a cluster algorithm with the grand tour can help refine partitions and the interpretation of the results.

More information on applying these methods can be found in Buja et al. (1988), Buja et al. (1991), Buja et al. (1996), Carr et al. (1996), Cook et al. (1996), Cook and Buja (1997), Cook (1997), Cook et al. (1997), Cook et al. (1998), and Majure et al. (1995).

4. Discussion

We have shown several simple examples of using interactive and dynamic graphics on atmospheric science data. These data are "gridded", but one of the important advantages of the tools described is that they will also work with non-gridded data. So while most methods for spatial data require gridding the data first, with these graphical tools it is possible to get some insight into spatial trends and dependencies without first creating a grid.

We also have a time aspect to our example data. The natural way to handle time is to strip it out as a separate variable, display of year in a second view, and allow the user to brush along the time values and watch the changes in the other plots. For example, to understand how the joint distribution of all four variables changes over years, we would setup a tour of sea surface temperature, sea level pressure, wind speed and wind direction in one view, and a dotplot display of year in a second view. Changes in the arrangement of views could allow us to examine seasonal trends over months, or individual monthly trends over years.

In summary, the paradigms that we adhere to in generating dynamic graphics are those of focusing, linking and arranging views, as described in Buja et al. (1996). These paradigms provide multiple, simple views such as scatterplots, in rearrangeable layouts, which have the ability to communicate information, such as brush color, to each other. With these simple approaches we can extract rather complex information from many varied types of data.

5. Acknowledgements

Márcia Macêdo's work was sponsored by CNPq—Brasil. Dianne Cook's work was funded in part by the U.S. Environmental Protection Agency through Cooperative Agreement CR822919 with Iowa State University. This paper has not been subjected to the Agency's peer and administrative review. No endorsement of the contents by the Agency should be inferred. The paper was revised while Dianne Cook was visiting the National Research Center for Statistics and the Environment, University of Washington, Seattle. Many thanks to Nicholas Lewin for help with the figures, and two referees and the editors for careful and detailed comments on this paper.

Notes

1. Details of the data can be found at: `http://ingrid.ldgo.columbia.edu/SOURCES/.COADS`, `http://ingrid.ldgo.columbia.edu/SOURCES/.COADS/.dataset_documentation.html`.
2. XGobi can be found at: `http://www.research.att.com/areas/stat/xgobi/index.html`.
3. *ArcView3.0* is a trademark of Environmental Systems Research Institute, Inc.
4. The ArcView-XGobi link can be found at: `http://www.public.iastate.edu/~arcview-xgobi/`.

References

Asimov, D. 1985. The grand tour: A tool for viewing multidimensional data. SIAM Journal of Scientific and Statistical Computing, 6(1):128–143.

Bao, S. and Anselin, L. 1997. Linking spatial statistics with GIS: Operational issues in the SPACESTAT-ARCVIEW link and the S+GRASSSLAND link. ASA Proceedings of the Section on Statistical Graphics: American Statistical Association, Alexandria, VA, pp. 61–66.

Buja, A. and Asimov, D. 1986. Grand tour methods: An outline. Proceedings of the 17th Symposium on the Interface between Computing Science and Statistics, D.M. Allen (Ed.). Lexington, KY: Elsevier, pp. 63–67.

Buja, A., Asimov, D., Hurley, C., and McDonald, J.A. 1988. Elements of a Viewing Pipeline for Data Analysis. In Dynamic Graphics for Statistics, W.S. Cleveland and M.E. McGill (Eds.). Monterery, CA: Wadsworth, pp. 277–308.

Buja, A., Cook, D., Asimov, D., and Hurley, C. 1997. Dynamic projections in high-dimensional visualization: Theory and computational methods. Technical report, AT&T Labs, Florham Park, NJ.

Buja, A., Cook, D., and Swayne, D. 1996. Interactive high-dimensional data visualization. Journal of Computational and Graphical Statistics, 5(1):78–99. See also `www.research.att.com/~andreas/xgobi/heidel/`.

Buja, A., McDonald, J.A., Michalak, J., and Stuetzle, W. 1991. Interactive data visualization using focusing and linking. Proceedings of Visualization '91, G.M. Nielson and L. Rosenblum (Eds.). IEEE Computer Society Press, Los Alamitos, CA, pp. 156–162.

Carr, D.B., Littlefield, R.J., Nicholson, W.L., and Littlefield, J.S. 1987. Scatterplot matrix techniques for large *N*. Journal of the American Statistical Association, 82:424–436.

Carr, D.B., Wegman, E.J., and Luo, Q. 1996. ExplorN: design considerations past and present. Technical Report 129, Center for Computational Statistics, George Mason University.

Cook, D. 1997. Calibrate your eyes to recognize high-dimensional shapes from their low-dimensional projections. Journal of Statistical Software, 2(6), `www.stat.ucla.edu/journals/jss/`.

Cook, D. and Buja, A. 1997. Manual controls for high-dimensional data projections. Journal of computational and Graphical Statistics, 6(4):464–480. Also see `www.public.iastate.edu/~dicook/research/papers/manip.html`.

Cook, D., Buja, A., Cabrera, J., and Hurley, C. 1995. Grand tour and projection pursuit. Journal of Computational and Graphical Statistics, 4(3):155–172.

Cook, D., Cruz-Neira, C., Kohlmeyer, B.D., Lechner, U., Lewin, N., Nelson, L., Olsen, A., Pierson, S. and Symanzik, J. 1998. Exploring environmental data in a highly immersive virtual reality environment. Environmental Monitoring and Assessment, 51(1–2):441–450. Also see `www.public.iastate.edu/~dicook/research/C2/statistic.html`.

Cook, D., Majure, J.J., Symanzik, J., and Cressie, N. 1996. Dynamic graphics in a GIS: Exploring and analyzing multivariate spatial data using linked software. Computational Statistics: Special Issue on Computer Aided Analyses of Spatial Data, 11(4):467–480.

Cook, D., Symanzik, J., Majure, J.J., and Cressie, N. 1997. Dynamic graphics in a GIS: More examples using linked software. Computers and Geosciences: Special Issue on Exploratory Cartographic Visualization, 23(4):371–385. `www.elsevier.nl/locate/cgvis`.

Dykes, J. 1996. Dynamic Maps for Spatial Science: A Unified Approach to Cartographic Visualization. In Innovations in GIS 3. London: Taylor & Francis, pp. 177–187.

Elms, J.D., Woodruff, S.D., Worley, S.J., and Hanson, C.S. 1993. Digitizing Historical Records for the Comprehensive Ocean-Atmosphere Data Set (COADS). Earth System Monitor, pp. 4–10.

Haining, R., Ma, J., and Wise, S. 1996. Design of a software system for interactive spatial statistical analysis linked to a GIS. Computational Statistics, 11(4):449–466.

Haslett, J., Bradley, R., Craig, P., Unwin, A., and Wills, G. 1991. Dynamic graphics for exploring spatial data with application to locating global and local anomalies. The American Statistician, 3:234–242.

MacDougall, E.B. 1992. Exploratory analysis, dynamic statistical visualization, and geographic information systems. Cartography and Geographic Information Systems, 19(4):237–246.

Macêdo, M. 1998. Exploratory Data Analysis of the Comprehensive Ocean-Atmosphere Data Set. MS Creative Component, Department of Statistics. Iowa State University, Ames, IA.

Majure, J.J., Cook, D., Cressie, N., Kaiser, M., Lahiri, S., and Symanzik, J. 1995. Spatial CDF Estimation and Visualization with Applications to Forest Health Monitoring. ASA Statistical Graphics Video Lending Library (contact: dfs@research.att.com).

McDonald, J.A. 1982. Interactive graphics for data analysis. Technical Report Orion II, Statistics Department. Stanford University, Stanford, CA.

McDonald, J.A. and Willis, S. 1987. Use of the Grand Tour in Remote Sensing. ASA Statiscal Graphics Video Lending Library (contact: dfs@research.att.com).

Monmonier, M. 1989. Geographic brushing: Enhancing exploratory analysis of the scatterplot matrix. Geographical Analysis, 21:81–84.

Newton, C. 1978. Graphica: From Alpha to Omega in Data Analysis. In Graphical Representation of Multivariate Data, P.C.C. Wang (Eds). New York: Academic Press, pp. 59–92.

Swayne, D.F., Cook, D., and Buja, A. 1998. XGobi: Interactive dynamic graphics in the X window system. Journal of Computational and Graphical Statistics, 71:113–130.

Symanzik, J., Klinke, S., Schmelzer, S., Cook, D., and Lewin, N. 1997. The ArcView/XGobi/XploRe environment: Technical details and applications for spatial data analysis. In ASA Proceedings of the Section on Statistical Graphics: American Statistical Association, Alexandria, Virginia. Forthcoming.

Symanzik, J., Majure, J.J., Cook, D., and Megretskaia, I. Linking ArcView 3.0 and XGobi: Insight behind the front end. Journal of Computational and Graphical Statistics, 1999. Forthcoming.

Unwin, A., Wills, G., and Hasslett, J. 1990. REGARD—Graphical analysis of regional data. ASA Proceedings of the Section on Statistical Graphics: American Statistical Association, Alexandria, VA, pp. 36–41.

Wilks, D.S. 1995. Statistical Methods in the Atmospheric Sciences. San Diego: Academic Press.

Woodruff, S.D., Lubker, S.J., Wolter, K., Worley, S.J., and Elms, J.D. 1993. Comprehensive Ocean-Atmosphere Data Set (COADS), Release 1a:1980-92'. Earth System Monitor, 1:1–8.

Márcia Macêdo is a graduate student at the Department of Statistics and Statistical Laboratory, Iowa State University. Her research interests include exploratory data analysis, visual data mining, spatial statistics.

Dianne Cook is an associate professor in the Department of Statistics and Statistical Laboratory at Iowa State University. She received her Ph.D. in Statistics from Rutgers University, the State University of New Jersey, funded by Bellcore, Morristown, New Jersey, in 1993. Her research interests include dynamic visualization of high-dimensional phenomena, exploratory data analysis, visual data mining, especially applied to geographical data.

Timothy J. Brown is a research scientist at the Desert Research Institute, and graduate faculty in the Atmospheric Sciences Program, University of Nevada. He received his Ph.D. in Climatology from the University of Colorado in 1995. His research interests include the application of statistical methods and visualization techniques to atmospheric science data.

www.ingramcontent.com/pod-product-compliance
Lightning Source LLC
LaVergne TN
LVHW011852200926
848484LV00002B/443
* 9 7 8 0 7 9 2 3 7 8 1 6 7 *